T0275966

SpringerBriefs in Applied Sciences and Technology

More information about this series at http://www.springer.com/series/8884

Moamar Sayed-Mouchaweh

Learning from Data Streams in Dynamic Environments

 Springer

Moamar Sayed-Mouchaweh
Computer Science and Automatic Control Department
High National Engineering School of Mines
Douai, France

ISSN 2191-530X ISSN 2191-5318 (electronic)
SpringerBriefs in Applied Sciences and Technology
ISBN 978-3-319-25665-8 ISBN 978-3-319-25667-2 (eBook)
DOI 10.1007/978-3-319-25667-2

Library of Congress Control Number: 2015956612

Springer Cham Heidelberg New York Dordrecht London

Printed on acid-free paper

Springer International Publishing AG Switzerland is part of Springer Science+Business Media
(www.springer.com)

Preface

This book treats the problem of learning from data streams generated by time-based and evolving nonstationary processes. It presents major and well-known techniques, methods, and tools able to manage, to exploit, and to interpret correctly the increasing amount of data in environments that are continuously changing. The goal is to build a predictor (classifier, learner), about the future system behavior, able to tackle and to govern the high variability of evolving and nonstationary systems. This book addresses the problems of modeling, prediction, classification, data understanding, and processing in nonstationary and unpredictable environments. It presents some major and well-known methods for the design of systems able to learn and to fully adapt their structure and to adjust their parameters according to the changes in their environments. In summary, this book aims at (1) defining the problem of learning from data streams in evolving and nonstationary environments, its interests, its applications, and its challenges, (2) providing a general scheme and principals of methods and techniques treating the problem of learning from data streams in evolving and nonstationary environments, (3) listing the major applications of these methods and techniques in various real-world problems, and (4) comparing these methods and techniques in order to define some new research directions in the area of learning from data streams in evolving and nonstationary environments.

Keywords: Drift monitoring; Self-adaptive learning methods; Data streams; Online learning; Incremental-decremental learning

Douai, France   Moamar Sayed-Mouchaweh

Contents

Chapter 1
Introduction to Learning

1.1 Introduction

The computerization of many life activities and the advances in data collection and storage technology lead to obtain mountains of data. They are collected to capture information about a phenomenon or a process behavior. These data are rarely of direct benefit. Thus, a set of techniques and tools are used to extract useful information for decision support, prediction, exploration, and understanding of phenomena governing the data sources. Machine learning (ML) [1] is the study, design, and development of methods and algorithms that enable computers to learn without being explicitly programmed. These methods and algorithms use historic data samples or observations about a process past behavior to build a predictor (classifier, recognizer, etc.). The latter is used as an old experience to predict the process future behavior of the system or the phenomenon generating the data.

Data mining (DM) [2] can be defined as the process of extracting knowledge or discovering unknown useful patterns automatically or semiautomatically, in large quantities of data. Therefore, DM may use ML in order to achieve intelligent tasks as humans in certain application domains.

Two main tasks of learning may appear in DM applications: classification and regression. In classification task, the learning aims at learning a way to classify unseen examples. Therefore, the output of classification learning is a discrete class label. In regression task, the learning looks for learning how to predict a numeric value of a variable. Therefore, in regression learning, the output is a numeric value of a variable. In both cases, the thing to be learned is called "concept," and the output produced by the learner is the description of this concept.

This chapter presents the basic definitions and notation related to the problem of learning from data samples. It shows how a learner (e.g., classifier) is built, and its performance is evaluated using multiple real and academic examples.

© The Author 2016   1
M. Sayed-Mouchaweh, *Learning from Data Streams in Dynamic Environments*,
SpringerBriefs in Applied Sciences and Technology,
DOI 10.1007/978-3-319-25667-2_1

Then, the limits of the assumptions used to build the learner are presented and illustrated by several examples. Finally, alternatives to overcome these limits are discussed.

1.2 Framework and Terminology

An instance, example, data point or observation $x \in \mathbb{R}^d$ that provides the input to a machine learning scheme is characterized by its values on a fixed, predefined set of d features or attributes (covariates). Therefore, $x = (x^1\, x^2 \ldots x^j \ldots x^d)$ can be seen as a point (vector) in a feature space of d dimensions formed by the different features or attributes. This space is called feature space, and x is called a pattern. The table gathering a set of n different instances $(x_1\, x_2 \ldots x_n)$ collected through n time steps $(t_1\, t_2 \ldots t_n)$, one instance at a time, is called the learning or training set X_L. The registered instances, one instance at a time, are not necessarily observed in equally spaced time intervals. In this table, the rows represent the instances, while the attributes or features are the columns. If X_L includes a set of pairs (x, y), labeled instances, where y is the corresponding output (discrete class label $\in \mathbb{Z}$ or numeric value $\in \mathbb{R}$), then the learning scheme is supervised, while if y is not provided, the learning scheme becomes unsupervised.

A learning model L (e.g., a classifier) attempts to predict the output y (e.g., class label) of the unseen incoming pattern $x = (x^1\, x^2 \ldots x^j \ldots x^d)$.

Example 1.1: Learning a Classifier Based on a Learning Set of Gaussian Distributions Consider the problem of classification of two Gaussian classes W_1 and W_2. W_1, respectively W_2, is described by one-dimensional normal distribution with the mean value $M_1 = 2$, respectively, mean value $M_2 = 7$, and the variance $\Sigma_1 = 1$, respectively, variance $\Sigma_2 = 1$, as follows:

$$x \in W_1 \Rightarrow x \sim N(M_1 = 2, \Sigma_1 = 1)$$
$$x \in W_2 \Rightarrow x \sim N(M_2 = 7, \Sigma_2 = 1)$$

The classifier L, learned based on a set of n_1 patterns from W_1 and n_2 patterns from W_2, is used to classify a new (unseen) pattern $x \in \mathbb{R}$ into one of these two classes.

Example 1.2: Learning How to Predict a Change (Up/Down) in the Electricity Price Let us take the problem of learning how to predict a change in the electricity price as a rise (up) and a fall (down) [3]. The concept to be learned is either a rise or a fall in the electricity price. The output target y is a binary class label: $y = 1$ if it is a rise (UP) and $y = 0$ if it is a fall (down). The input vector $x = (x^1, x^2, x^3)$ comprises three numerical features or covariates. x^1 is the electricity demand in the studied area (region or state), x^2 is the electricity demand in the adjacent area, and x^3 represents the scheduled electricity transfer between these two areas. Based on x at the time step t, the learner L must predict if the electricity price will go up

($y = 1$) or down ($y = 0$). Therefore, the learner will be a classifier. The latter is learned based on the use of a set X_L of historical records in the past about the values of x^1, x^2, and x^3.

Example 1.3: Learning How to Detect a Fraudulent Transaction In a fraud detection application [4], the output target y is a binary class label: $y = 1$ if a given transaction is fraudulent and 0 if it is genuine. Every credit card transaction x consists of features (variables, covariates) describing a behavioral characteristic of the card holder usage. These features show the spending habits of the customers with respect to geographical locations, days of the month, hours of the day, and merchant category codes representing the type of the merchant where the transaction takes place. The set X_L of historical transitions of a credit card is used to learn a model or a classifier L used in the fraud detection systems to distinguish fraudulent activities from genuine ones. The fraudulent activities can be seen as a significant deviation from the usage profile of the credit card holder representing the genuine class.

Example 1.4: Learning How to Detect a Spam A spam filter [5] can be designed as a classifier L able to assign an email message in one of two classes: spam ($y = 1$) or legitimate ($y = 0$). An instance x representing a mail is characterized by several numeric features as the sending time or the number of receptors and textual features describing the mail contain. The textual features capture the presence of words in the body of an email relevant to spam as the word "Viagra." A classifier, or filter, L is built based on the use of a set X_L of historical mails.

Example 1.5: Learning How to Distinguish Between Six Different Gases Electronic nose [6] for chemical analysis is based on the use of a set of 16 chemical (gas) sensors. The training or leaning set X_L is formed by gathering the sensors measurements during 36 months [7]. The classification task is to distinguish between six classes representing six different gases dosed at different concentrations. The classifier L is learned based on the use of X_L. For a new pattern x gathering the current measurements of six chemical or gas sensors, L assigns x to one of the six different classes in order to determine the gas kind and concentration.

1.3   Learner Design

The goal of the learner design is to build a model L (e.g., classifier or predictor) able to predict the class label y_i ($i = 1, 2, \ldots, c$) of an incoming unseen instance x. To accomplish that, the distribution D, which represents the joint probability $P(y, x)$, must be estimated:

$$D = \{P(y_1, x), P(y_2, x), \ldots, P(y_c, x)\} \tag{1.1}$$

Therefore, D defines the concept to be learned.

$P(y_i, x)$, $(i = 1, 2, \ldots, c)$, can be calculated at any time step based on the use of the marginal probability $P(y_i)$, indicating the prior probability of y_i in the sense that it does not take into account any information about x, and the conditional probability $P(x|y_i)$ as follows:

$$P(y_i, x) = P(x|y_i)P(y_i) \qquad (1.2)$$

Similarly, $P(y_i, x)$ can be calculated as follows:

$$P(y_i, x) = P(y_i|x)P(x) \qquad (1.3)$$

The posterior probability $P(y_i|x)$ quantifying the probability that x belongs to the class W_i can be calculated using the Bayesian decision theorem [8] as follows:

$$P(y_i|x) = \frac{P(x|y_i)P(y_i)}{P(x)} \qquad (1.4)$$

$P(x)$ denotes the marginal or prior probability of x and it acts as a normalizing constant. It represents the evidence factor that is used as a scale factor in order to guarantee the sum of posterior probabilities for x is equal to one:

$$P(y_1|x) + \ldots + P(y_i|x) + \ldots + P(y_c|x) = 1 \qquad (1.5)$$

By substituting (1.4) in (1.5), $P(x)$ is calculated as follows:

$$P(x) = P(x|y_1)P(y_1) + \ldots + P(x|y_i)P(y_i) + \ldots + P(x|y_c)P(y_c) \qquad (1.6)$$

To make a classification decision, determining the class label y_i, for unseen x in the case of equal costs of mistake, the classifier L assign x to the class for which it has the maximal posterior probability:

$$y(x) = y_i : i \in \{1, 2, \ldots, c\} \Rightarrow P(y_i|x) = Max\left(P\left(y_j|x\right)\right), \forall j \in \{1, \ldots, i, \ldots, c\} \qquad (1.7)$$

Example 1.6: Design a Classifier Based on Gaussian Distributions Let us take Example 1.1, representing two Gaussian classes described in one-dimensional feature space, and let us design a classifier L to assign a new pattern x to one of these two classes. Figure 1.1 shows the conditional probability $P(x|y_1)$, respectively, $P(x|y_2)$, in class W_1, respectively, class W_2. The Gaussian probability can be calculated by [8]

$$P(x|y_i) = (2\pi)^{-d/2} \times |\Sigma_i|^{-\frac{1}{2}} \times \exp\left[-0, 5 \times (x - M_i)^{tr} \times \Sigma_i^{-1} \times (x - M_i)\right] \qquad (1.8)$$

Where d is the dimension of the feature space which is equal to 1 for this example.

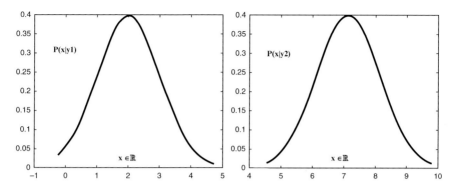

Fig. 1.1 Conditional probability $P(x|y_1)$, respectively $P(x|y_2)$, in class W_1, respectively class W_2

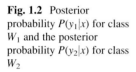

Fig. 1.2 Posterior probability $P(y_1|x)$ for class W_1 and the posterior probability $P(y_2|x)$ for class W_2

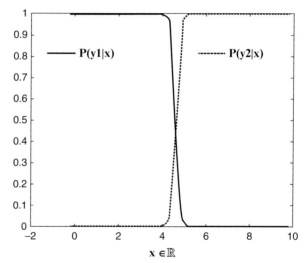

The posterior probability $P(y_1|x)$ for W_1 and the posterior probability $P(y_2|x)$ for W_2 are calculated based on the use of (1.4). They are depicted in Fig. 1.2.

The decision boundary is calculated using (1.7). It divides the feature space into two zones. All the patterns located in the first zone will be classified in class W_1 ($y = 1$), and the ones located in the second zone will be classified in W_2 ($y = 2$). Therefore, the decision boundary allows the classifier L to make a decision on the class of a new incoming pattern x. Figure 1.3 shows this boundary decision in the one-dimensional feature space.

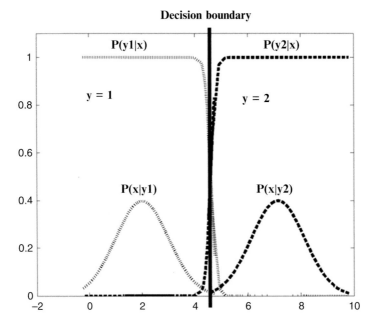

Fig. 1.3 Decision boundary of the two Gaussian classes in one-dimensional feature space

Table 1.1 Two-class evaluation outcome

	Predicted as positive	Predicted as negative
True class: positive	TP (true positive)	FP (false positive)
True class: negative	FN (false negative)	TN (true negative)

1.4   Performance Evaluation

When a big amount of data is available, the model (e.g., classifier or predictor) can be built using a large training set, and its performance can be evaluated using another large test set. In general, three datasets are used: the *training or learning* data, the *validation* data, and the *test* data. The training data set is used to learn the classifier. The validation data set is used to optimize its parameters. Then the test data set is used to calculate its performance. Each of the three data sets must be chosen independently. It is assumed that both the training data and the test data sets include representative samples of the underlying problem or concept.

   In order to evaluate the performance of the designed classifier, the following evaluation metrics are applied using the test data set for the case of two classes (see Table 1.1):

- Accuracy (Acc) or error (Err) of prediction:

$$\text{Acc} = \frac{\text{TP} + \text{TN}}{\text{TP} + \text{TN} + \text{FP} + \text{FN}} = 1 - \frac{\text{FP} + \text{FN}}{\text{TP} + \text{TN} + \text{FP} + \text{FN}} = 1 - \text{Err} \quad (1.9)$$

- Specificity (Spe):

$$\text{Spe} = \frac{\text{TP}}{\text{TP} + \text{FN}} \quad (1.10)$$

- Sensitivity or recall (Sen):

$$\text{Sen} = \frac{\text{TN}}{\text{TN} + \text{TP}} \quad (1.11)$$

- Precision (Pre):

$$\text{Pre} = \frac{\text{TP}}{\text{TP} + \text{TN}} \quad (1.12)$$

In the two-class case (positive, negative), a single prediction or classification has the four different possible outcomes shown in Table 1.1. The *true positives* (TP) and *true negatives* (TN) are correct classifications. A *false positive* (FP) denotes the case where the pattern is incorrectly predicted as positive while it is actually *negative*. A *false negative* (FN) indicates the case where the pattern is incorrectly predicted as negative when it is actually positive.

An overall performance indicator can be a combination of the previous evaluation metrics as weighted prediction accuracy:

$$\text{WPA} = \frac{(w_1 \times \text{Acc}) + (w_2 \times \text{Spe}) + (w_3 \times \text{Sen}) + (w_4 \times \text{Pre})}{w_1 + w_2 + w_3 + w_4} \quad (1.13)$$

$\text{TP} + \text{TN} + \text{FP} + \text{FN}$ is the number of patterns in the test data sets. w_1, w_2, and w_3 are weights used to determine the importance of contribution of each evaluation criterion to the prediction accuracy of the designed model (classifier). These weights are used in order to take into account the effect of imbalanced data when the number of patterns in one class is much larger than the one in the other class.

In multiclass prediction, the evaluation outcome is often displayed as a two-dimensional *confusion matrix* with a row and column for each class. Each matrix element shows the number of test examples for which the actual class is the row and the predicted class is the column.

It is worth to note that the accepted prediction accuracy depends on the application context and its conditions.

Example 1.7: Spam Filter (classifier) Evaluation The spam filter (see Example 1.4) aims at classifying a mail as spam or legitimate. A serious problem for spam filters is the occurrence of false positives (FPs), i.e., legitimate emails classified incorrectly as spam. FP is significantly more serious than false negative (FN) (a spam email incorrectly classified as legitimate). For many people, the occurrence of FPs is unacceptable. One way to bias the spam classifier away from FPs, the classifier used unanimous voting to determine whether the mail was spam or not [9]. All neighbors returned had to have a classification decision of the mail as spam in order to be classified as spam.

1.5   From Static Toward Dynamic Environments

In traditional machine learning and data mining approaches, the current observed data and the future data are assumed to be sampled independently and from an identical probability distribution (iid). The assumption of independence means that the data samples, generated over time by a variable characterizing a phenomenon, are statistically independent. Therefore, past and current data samples do not affect the probability for future ones. The assumption of identically distributed observations means that generated observations over time may be considered as random draws from the same probability distribution. Let x be an observation and y its true label, and let $(P_t(x,y))$ be the joint probability that presents the concept at time t. Identical distribution means that the joint probability of an observation and its label is the same at any time $P_{t_1}(x, y) = P_{t_2}(x, y)$. Independent distribution means that the probability of a label does not depend on what was observed earlier $P(y_t) = P(y_t|y_{t-1})$.

   However, in multiple applications like web mining, social networks, network monitoring, sensor networks, telecommunications, financial forecasting, etc., data samples arrive continuously online through unlimited streams often at high speed, over time. Moreover, the phenomena generating these data streams may evolve over time. In this case, the environment in which the system or the phenomenon generated the data is considered to by dynamic, evolving or nonstationary.

   Therefore, machine learning and data mining methods used to learn from data generated from phenomena evolving in dynamic environments are facing three challenges. The first is caused by the huge amount of data that are continuously generated over time. The second challenge is due to the high speed of arrival of data streams. Finally, the third challenge may occur when the joint probability distribution of the data samples evolve over time. Hence, data generated by phenomena in dynamic environments are characterized by: (a) potentially unlimited size; (b) sequential access to data samples in the sense that once an observation has been processed, it cannot be retrieved easily unless it is explicitly stored in memory; and (c) unpredictable, dependent, and not identical distributed observations.

   Consequently, learning from streams of evolving and unbounded data requires developing new algorithms and methods able to learn under the following constraints:

- Random access to observations is not feasible, or it has high costs. This means that the entire original dataset is not a priori available or it is too large to process.
- Memory is small with respect to the size of data.
- Data distribution or phenomena generating the data may evolve over time. This is also known as concept drift.

Example 1.8: Classifier of the Electricity Price in Dynamic Environments Example 1.2 presented a classifier to predict a change in the electricity price as a rise (up) and a fall (down). The data samples represent the recorded values of the electricity demand in the studied area or region, the electricity demand in the adjacent area, and the scheduled electricity transfer between these two areas. In this electricity market, there are two aspects to be taken into account:

- The input vector x is recorded each 30 min. Therefore, data arrives as streams of records each day.
- The prices are not fixed and may be impacted by the demand and supply.

Therefore, the data streams are subject to changes over time due to changes in consumption habits, to unexpected events (severe winter, hot summer, political problems, etc.) and seasonality or to the expansion of the electricity market to include adjacent areas. This allows for the production surplus of one region to be sold in the adjacent region, which in turn dampened the electricity price. Therefore, data samples are subject to concept drift due to changing consumption habits, unexpected events, and seasonality. Hence, the classifier needs to be updated overtime to take into account these changes.

Example 1.9: Classifier of Fraudulent Transactions in Dynamic Environments As we have seen in Example 1.3, the output target y is a binary class label: $y = 1$ if a given transaction is fraudulent and 0 if it is genuine. The transitions (data samples) arrive continuously over time. Therefore, data loads are huge. In addition, the decision (transaction fraudulent or genuine) needs to be taken fast, almost online, to stop the crime. Therefore, the performance of a static classifier that was previously trained using a fixed learning or training set is bound to degrade over time because:

- Only a small set of supervised samples is provided by human investigators who have time to assess only a reduced number of alerts. Labels of the vast majority of transactions are made available only several days later, when customers have possibly reported unauthorized transactions.
- The distribution of genuine and fraudulent transactions evolves over time because of seasonality (e.g., customers' behavior change in holiday seasons) and new attack strategies (new fraud activities may appear).

Example 1.10: Spam Classifier in Dynamic Environments As we have seen in Example 1.4, the spam classifier aims to predict if a mail is a spam or legitimate. The mails arrive as streams over time. The spam classifier detects a spam based on a set of features capturing the presence of certain words (e.g., Viagra) in the body of an email. Therefore, a spam classifier based on static textual and contextual features may become unusable very quickly. This is due to the change of the user interests over time; the words relevant to spam for an individual change at different periods of time. In addition, new words can better discriminate the user new interests or the spam since new unsolicited commercials come into vogue. This is known as the problem of hidden context features. Moreover, the spammers alter, obfuscate, and confuse filters by disguising their emails to look more like legitimate email by replacing letters with numbers or by inserting irrelevant characters between letters (e.g., Viagra becomes v1agra or v.i.a.g.r.a, etc.). Consequently, the dynamic nature of spam emails requires a continuous update of the features used by the spam classifier (filter) in order to distinguish correctly between spam and legitimate mails.

Example 1.11: Classifier of Different Gases in Dynamic Environments Example 1.5 presented a classifier (electronic nose) to distinguish between six gases dosed at different concentrations. The gradual and unpredictable change of sensor responses, so-called sensor drift, is one of the major challenges to develop electronic nose. This drift can be defined as a slow, in the case of long-term drift, or abrupt change of the measured chemical property independently of its real value. The sensor drift can be due to the aging and poisoning or to the measurement conditions of the sensors environments as the humidity and temperature. The sensor drift impacts the performance of the classifier. Therefore, it is important to update the classifier in response to the drift in the sensors in order to compensate its consequences on the measured chemical properties of gases.

Consequently in order to answer the challenges presented in these examples, the learner (e.g., classifier) must be able to achieve:

- Online learning in order to learn from unlimited size of data streams
- Incremental learning in order to integrate the information carried by each new arriving data sample with limited time and memory
- Decremental learning in order to forget or unlearn the data samples which are no more useful (obsolete or outdated)

Therefore, online and self-adaptive machine learning and data mining approaches need to be developed in order to generate models (classifiers, predictors) that can adapt their parameters and structures online according to the changes and novelties carried out by the new arriving data streams.

Chapter 2
Learning in Dynamic Environments

2.1 Introduction

The volume of data is rapidly increasing due to the development of the technology of information and communication. This data comes in the form of streams. Learning from this ever-growing amount of data requires a continuous learning over time. Traditional one-shot memory-based learning methods trained offline from a fixed size of historic data set are not adapted to learn from these data streams. This is because firstly, it is not feasible to register all the data samples over time, and secondly the generated models become quickly obsolete due to the occurrence of changes, also known as "concept drift," in their environments.

Therefore, as we have seen in Chap. 1, online self-adaptive learning scheme is required in order to accommodate the new information carried out by the new incoming data samples and to unlearn the obsolete or outdated ones due to the changes in the learner's or model's environments.

In this chapter, the problem of drifting data streams in dynamic environments is formalized, and its framework is defined. Then, the kinds and characteristics of the concept drift are presented. Finally, the real-world applications generating drifting data streams are discussed. The goal is to give a picture of the problem of learning from data streams in dynamic environments, its causes, sources, and characteristics in order to discuss later alternatives to solve this problem.

© The Author 2016
M. Sayed-Mouchaweh, *Learning from Data Streams in Dynamic Environments*,
SpringerBriefs in Applied Sciences and Technology,
DOI 10.1007/978-3-319-25667-2_2

2.2 Concept Drift Framework

2.2.1 Incremental Learning

Let X_L be a set of historic labeled data samples available at time t. It contains n labeled samples: (x_i, y), $i = 1, .., n, y \in \{1, \ldots, c\}$, collected in the past until the time t and distributed into c classes. Let L_t be the learner (e.g., classifier) built using X_L. L_t is used to predict the class label y_{t+1} for a new unseen data sample x_{t+1}. Since after this classification the label y_{t+1} is available, then the learner L_t can be updated by integrating (x_{t+1}, y_{t+1}) to its learning or training set. If the update of L_t is achieved by using only the new labeled data sample, and not the whole X_L, then the learning is called incremental learning [10–12].

As we have seen in Chap. 1, a concept or a source S can be defined as the classes' joint probabilities $\{P(y_1, x), P(y_2, x), \ldots, P(y_c, x)\}$ which in their turn are defined by the classes' prior probabilities $\{P(y_1), P(y_2), \ldots, P(y_c)\}$ and the classes' conditional probabilities $\{P(y_1|x), P(y_2|x), \ldots, P(y_c|x)\}$ (see (1.2)):

$$S = \{(P(y_1), P(x|y_1)), (P(y_2), P(x|y_2)), .., (P(y_c), P(x|y_c))\} \qquad (2.1)$$

Every data sample x_i is generated by the source S_i. Therefore, if $S_1 = S_2 = \ldots = S_n = S$, then the concept, defined by S or D, is stable. The incremental model can improve its performance by approximating D as the number of incoming data increases. Thus, the hypothesis (*concept*) learned before are still valid and can be efficiently approximated whenever the number of data samples increases to infinite. This is known by stable or stationary concept.

Consequently, incremental learning [10–12] is a suitable online learning scheme allowing to learn from infinite streams of data samples using limited time and memory size. Therefore, incremental learning methods do not require the availability of an initial complete training set since they continue to learn from the incoming data samples over time. However, they assume that the hypotheses (source, concept, distribution, etc.) learned before are always valid for the new incoming data. This reduces the ability of incremental learning methods to evolve the model (predictor, classifier) at the same rate as data streams.

In this context, they are considered to be suitable for learning from data streams since there is no control on their order of arrive nor their representativeness (i.e., the data samples may not be independent, and they may not be randomly generated from the source S). Moreover, they are theoretically able to continuously improve their performance as the number of incoming data increases. However, this can be true *if and only if* we assume that the hypotheses (S or D) generating the data streams are the same over time.

Example 2.1: Comparison Between Static and Incremental Classifiers Let us consider two classes described in one-dimensional feature space as follows:

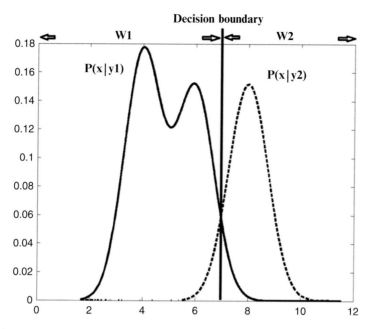

Fig. 2.1 Conditional probability densities for Gaussian classes W_1 and W_2 and the corresponding optimal boundary decision learned from a complete learning set

$$x \in W_1 \Rightarrow x \sim N(M_1 = 4, \Sigma_1 = 1) + N(M_1 = 6, \Sigma_1 = 1)$$
$$x \in W_2 \Rightarrow x \sim N(M_2 = 8, \Sigma_2 = 1)$$

We suppose that both classes have the same prior probability: $P(y_1) = P(y_2)$. We can see that the data samples of W_1 belong to two different Gaussian distributions. Figure 2.1 presents the conditional probabilities $P(x|y_1)$ and $P(x|y_2)$ according to classes W_1 and W_2 as well as the optimal boundary decision. Let us suppose that the available data samples belonging to W_1 in X_L were generated only from $x \sim N(M_1 = 4, \Sigma_1 = 1)$. Therefore, the learner will learn the conditional probabilities $P(x|y_1)$ and $P(x|y_2)$ depicted in Fig. 2.2. If the classifier is static, the new incoming data samples will be classified according to the decision boundary depicted in Fig. 2.2. Based on (1.7), the patterns between the expected decision boundary of Fig. 2.1 and the one learned by the static classifier will be misclassified since they will be classified in W_2 while they are in W_1. An incremental classifier will update its decision boundary according to the incoming new data samples. The ones of the latter generated by $x \sim N(M_1 = 6, \Sigma_1 = 1)$ will allow the incremental classifier to improve its performance (accuracy or classification rate) by updating its decision boundary (see Fig. 2.1).

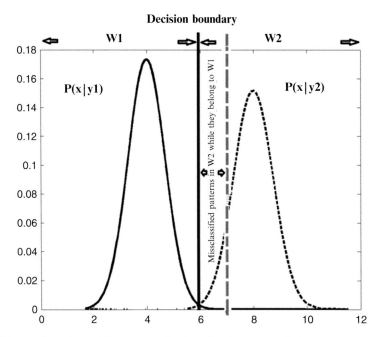

Fig. 2.2 Conditional probability densities for Gaussian classes W_1 and W_2 and the corresponding optimal decision boundary learned from an incomplete learning set. The *gray dashed* decision boundary represents the boundary decision in the case of complete learning set. All the data samples between these two decision boundaries will be misclassified in W_2 while they belong to W_1

2.2.2 Adaptive Learning

One major challenge arises when the underlying source generating the data is not stationary: $S_1 \neq S_2 \neq \ldots \neq S_n$. This leads to a change in the data distribution according to a single feature, to a combination of features or in the class boundaries. This is known as concept drift. In this case, the assumption of data identically distributed is no more satisfied, and the incremental learning is no longer able to approximate the distribution of the new incoming data samples. Indeed, incremental learning considers the already learned concepts are valid. This is the case in many real-world domains where the concept of interest may depend on some hidden context, not given explicitly in the form of predictive features.

Example 2.2: Comparison Between Static, Incremental, and Adaptive Classifiers Let us consider two Gaussian classes described in one-dimensional feature space as follows:

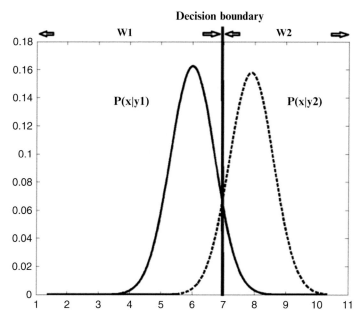

Fig. 2.3 Conditional probability densities $P(x|y_1)$ and $P(x|y_2)$ for Gaussian classes W_1 and W_2 and the corresponding optimal decision boundary learned before the concept drift in W_1

$$x \in W_1 \Rightarrow x \sim N(M_2 = 6, \Sigma_2 = 1)$$
$$x \in W_2 \Rightarrow x \sim N(M_1 = 8, \Sigma_1 = 1)$$

We suppose that both classes have the same prior probability: $P(y_1) = P(y_2)$. Figure 2.3 presents the conditional probabilities $P(x|y_1)$ and $P(x|y_2)$ according to classes W_1 and W_2 as well as their optimal boundary decision. Let us now suppose that the new incoming data samples do not follow any more $x \sim N(M_1 = 6, \Sigma_1 = 1)$ but are generated by new source or distribution defined by $x \sim N(M_1 = 4, \Sigma_1 = 1)$. Therefore, W_1 will be defined by the new conditional probability density depicted in Fig. 2.4. Due to this concept drift in W_1 ($x \sim N (M_1 = 6, \Sigma_1 = 1) \rightarrow x \sim N(M_1 = 4, \Sigma_1 = 1)$), the decision boundary must be updated as it is depicted in Fig. 2.4. However, the incremental learning cannot update correctly the boundary decision since it considers all the data samples, and therefore their sources or distributions, are valid (see Fig. 2.5).

Consequently, the data samples generated by the old distribution or source of W_1, $x \sim N(M_1 = 6, \Sigma_1 = 1)$, must be removed or unlearned, and only the data samples generated by the new source or distribution ($x \sim N(M_1 = 4, \Sigma_1 = 1)$) must be used to update the decision boundary.

In order to achieve the most accurate classification or prediction in the presence of concept drift, the learner should be able to track such drift and quickly adapt to it. Therefore, self-adaptive online learning [13–16] is the most adequate learning

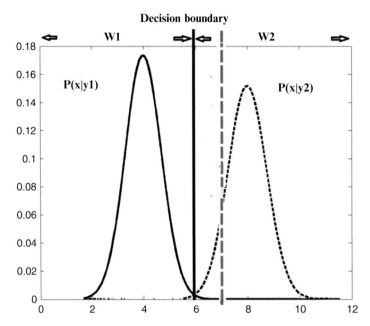

Fig. 2.4 Conditional probability densities $P(x|y_1)$ and $P(x|y_2)$ for Gaussian classes W_1 and W_2 and the corresponding optimal decision boundary learned using the data samples generated by the new source of W_2 and removing or unlearning the ones generated by the old source. The *gray dashed* boundary decision represents the initial boundary decision before the concept drift in W_1

scheme to learn from evolving data streams in dynamic environments since they integrate a forgetting mechanism of outdated or obsolete data. To achieve that, two questions must be answered: (1) how to track concept drift and (2) how to adapt the learner parameters and structure in order to react to this concept drift. To answer these questions, the drift causes, types, and characteristics will be detailed in the next. This will allow defining how data samples that are representative of the new concept (the new source or distribution generating the data) can be determined and how they can be used to adapt the classifier parameters and structure.

2.3   Causes and Kinds of a Concept Drift

Based on (1.4) and (1.7), there are FOUR terms that are used by the Bayes formula to achieve the classification task:

- The prior probability $P(y_i)$ of each class W_i
- Its conditional probability $P(x|y_i)$
- The posterior probability $P(y_i|x)$
- The marginal or prior probability $P(x)$ of x

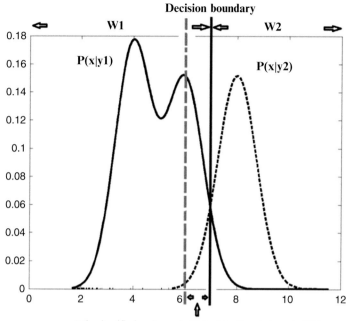

Fig. 2.5 Conditional probability densities $P(x|y_1)$ and $P(x|y_2)$ for Gaussian classes W_1 and W_2 and the corresponding optimal decision boundary learned using all the data samples generated by the old and new concepts of W_1. The *gray dashed line* represents the decision boundary learned by using only the data samples generated from the new source. All the data samples between these two decision boundaries will be misclassified in W_2 while they belong to W_1

$P(x)$ is constant for all the classes since it acts as a normalization or evidence factor. Therefore, a concept drift may result due to a change in:

- The prior probability $P(y_i)$ of a class W_i
- The conditional probability $P(x|y_i)$ in a class W_i
- The posterior probability $P(y_i|x)$ of a class W_i
- A combination of them

It is worth mentioning that a change in the class prior probability leads to class imbalance, novel class emergence, existing classes' fusion, or existing classes' splitting.

These changes can cause two kinds of concept drift: real and virtual. The real concept drift refers to changes in the classes' posterior probabilities $P(y_i|x)$. This means that the target concept, y, for a pattern x with the same values of features will change in response to the occurrence of a drift. Therefore, this kind of drift directly impacts the decision boundary, which in turn decreases the performance of the learner if the latter is not updated. The virtual concept drift refers to changes in the classes' conditional probabilities $P(x|y_i)$ without impacting the posterior probabilities $P(y_i|x)$. Therefore, this kind of concept drift impacts the data distribution within the same class in the feature space without affecting the corresponding decision boundaries.

Example 2.3: Real drift caused by a change in the classes' prior probabilities Let us take the case of two Gaussian classes described in one-dimensional space as follows:

$$x \in W_1 \Rightarrow x \sim N(M_1 = 2, \Sigma_1 = 1)$$
$$x \in W_2 \Rightarrow x \sim N(M_2 = 5, \Sigma_2 = 1)$$

Let us suppose that W_1 represents the healthy human subjects, while W_2 gathers the ill ones. Firstly, we suppose that the learning set X_L includes 200 data samples: 160 data samples belong to class W_1, and the other 40 ones belong to class W_2. Therefore, the prior probabilities for W_1 and W_2 are: $P(y_1) = 0.8$ and $P(y_2) = 0.2$. $P(y_1)$ is significantly bigger than $P(y_2)$ because in normal conditions, we expect more healthy human subjects than ill ones.

Let us suppose that within the next 800 inspected human subjects: 40 were assigned to W_1, while 760 ones assigned to W_2. This means that $P(y_1)$ will be significantly decreased to 0.2, while $P(y_2)$ will be increased to 0.8. This significant change in the prior probabilities of both classes indicates a change (drift) in the classifier environments. This change can be due to an epidemic spread. The latter increases significantly the number of contaminated human subjects (e.g., W_2) and reduces the number of healthy human subjects (e.g., W_1) according to the total number of inspected human subjects. This change entails a change in the posterior probabilities of both classes as well as in their optimal decision boundary as we can see in Fig. 2.6. If the original decision boundary is not updated in response to this change, all the patterns located between the original and new decision boundaries will be misclassified. This leads to decrease the classifier performance (the rate of patterns correctly classified).

We must mention here that the conditional probabilities $P(x|y_1)$ and $P(x|y_2)$ of classes W_1 and W_2 did not change; they are the same before and after the change in the classes' prior probabilities.

Example 2.4: Real Concept Drift Caused by a Change in the Classes' Posterior Probabilities Let us take the example of the classifier of the electricity price tendency, treated in Examples 1.2 and 1.8. Let us suppose that the prices of oil have been changed (decreased or increased) because of a political or economic crisis event. Then for the same values of the input vector $x = (x^1, x^2, x^3)$, representing the electricity demand in the studied area, the electricity demand in the adjacent area, and the scheduled electricity transfer between these two areas, the estimated class (up/down) may become different from the one estimated without the occurrence of this unpredicted event. If this change or drift is not taken into account, the classifier prediction accuracy will be decreased.

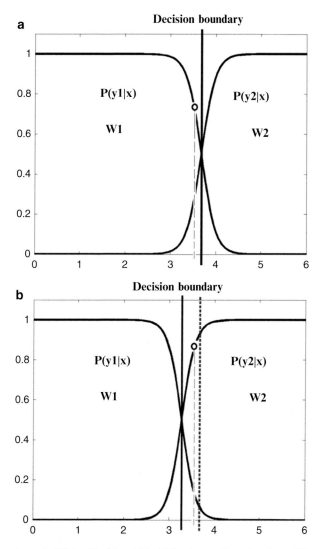

Fig. 2.6 Posterior probabilities $P(y_1|x)$ and $P(y_2|x)$ for the two Gaussian classes W_1 and W_2 as well as their corresponding decision boundary before (**a**) and after (**b**) the change in the classes' prior probabilities $P(y_1)$ and $P(y_2)$. The *dashed gray line* in (**b**) shows the decision boundary before the change in the prior probabilities $P(y_1)$ and $P(y_2)$ of W_1 and W_2. For the round point representing a data sample, it belongs to W_1 before the drift (**a**), and after the drift it belongs to W_2 (**b**). We can see clearly that the decision boundary and posterior probabilities have been changed in response to this change in the prior probabilities of both classes

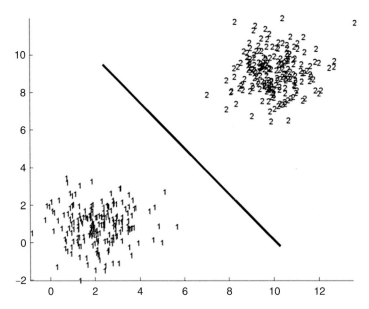

Fig. 2.7 Two Gaussian classes and their corresponding decision boundary in two-dimensional feature space

Example 2.5: Virtual Concept Drift Caused by a Change in the Classes' Conditional Probabilities Let us take the case of two Gaussian classes described in two-dimensional space as follows:

$$x \in W_1 \Rightarrow x \sim N(M_1 = (2, 1), \Sigma_1 = (1, 1))$$
$$x \in W_2 \Rightarrow x \sim N(M_2 = (10, 9), \Sigma_2 = (1, 1))$$

Figure 2.7 depicts these two classes in the feature space as well as their corresponding decision boundary. Let us suppose that a drift has occurred in the class W_1 leading to a change in its mean values as follows:

$$x \in W_1 \Rightarrow x \sim N(M_1 = (3, 3), \Sigma_1 = (1, 1))$$

This change in the parameters (mean values) of the normal law generating the data samples is represented as a change (a move) of the location of W_1 in the feature space (see Fig. 2.8). This drift changes the data spatial distribution of W_1 in the feature space without impacting the decision boundary. Indeed, the performance of the classifier with the initial decision boundary (see Fig. 2.7) will not be impacted by this drift in W_1 since both classes after the drift remain perfectly separated by the initial decision boundary (see Fig. 2.8). Therefore, the drift in the class W_1 is virtual since it does not impact the classifier performance.

Let W_1 and W_2 be, respectively, the normal and failure operation conditions of a machine. The classifier aims at assigning an incoming pattern representing the

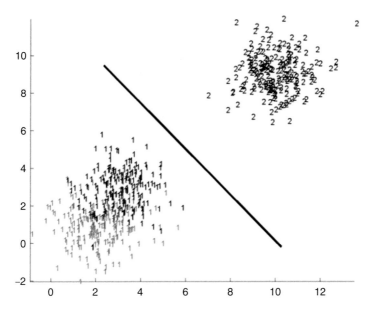

Fig. 2.8 Virtual drift in the class W_1 leading to change its spatial location or distribution in the feature space. The *gray points* of W_1 indicate the initial patterns belonging to W_1 before the drift. This drift does not impact the decision boundary since the initial one still able to separate perfectly both classes after the drift in W_1

machine's current operation conditions into normal or failure classes. When this machine starts to malfunction, its performance to accomplish a task decreases over time. However, as long as this performance remains acceptable (greater than the threshold defined for the failure representing unacceptable decrease in the machine's performance), the classifier continues to classify properly the incoming patterns as belonging to the class of normal operation conditions although the characteristics of the normal class are drifting over time. Therefore, in this case, the concept drift is virtual.

Example 2.6: Virtual Concept Drift Becoming Real Concept Drift Let us take Example 2.5 where the conditional probability $P(x|y_1)$ of class W_1 changed as follows:

Before drift: $x \in W_1 \Rightarrow x \sim N(M_1 = (2, 1), \Sigma_1 = (1, 1))$.
After drift: $x \in W_1 \Rightarrow x \sim N(M_1 = (3, 3), \Sigma_1 = (1, 1))$.

Let us suppose now that a new concept drift occurred in the conditional probability $P(x|y_1)$ of W_1 as follows (see Fig. 2.9):

$$x \in W_1 \Rightarrow x \sim N(M_1 = (5, 6), \Sigma_1 = (1, 1))$$

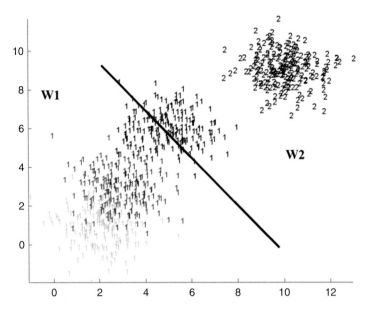

Fig. 2.9 Drift in the class W_1 passing from virtual to real concept drift. When the drift in W_1 becomes real, the decision boundary is adversary impacted since the patterns of W_1 enter the area of W_2 in the feature space. All the patterns of W_1 located in the area of W_2 in the feature space will be then misclassified

Some of the data samples belonging to the new concept drift of W_1 will occupy the zone of class W_2 in the feature space. Therefore, these data samples will be misclassified by the static classifier since the latter will assign them to W_2 while they belong to W_1. Consequently, the classifier performance (accuracy) will be decreased if this drift is not taken into account in order to update the decision boundary of the classifier.

Let us suppose that W_1 represents the class of spam while W_2 is the class gathering the legitimate emails. If the spammer looks to trick the spam filter (the classifier) by changing its behavior in order to be similar to the one of legitimate emails (class W_2), then the spatial position of W_1 in the feature space will be moved in order to occupy the zone of W_2. When the patterns of W_1 enter the zone in the feature space that is considered by the classifier (spam filter) as the behavior of legitimate mails (W_2), then the spam emails will be considered as legitimate. This will lead to decrease the performance of the spam filter. Therefore, the decision boundary of the classifier (spam filter) must be updated in order to take into account the change in the behavior of the spam emails induced by the spammer to trick the spam filter. Figure 2.10 shows the new decision boundary that allows maintaining the classifier performance for the example of Fig. 2.9.

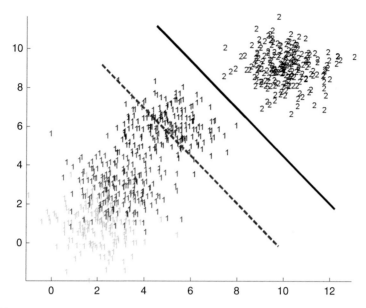

Fig. 2.10 Updating the decision boundary of the classifier in order to maintain its performance when the drift in the class W_1 passed from virtual to real concept drift. The *solid line* is the updated decision boundary of the initial one (in *gray dashed line*) before the drift

2.4   Concept Drift Description

A drift can be represented according to different criteria which are used to describe how the new concept replaces the old one [17–19]. These criteria give indications about the drift period, its speed, its intensity or severity, its frequency, and whether it can be detected or not. These characteristics are very useful to guide the choice and to define the framework of the methods and tools suitable to handle concept drift.

2.4.1   Drift Speed

The drift duration, called also drifting time or drift width, is the number of time steps for a new concept to replace the old one in the sense that no data samples of the old samples will occur. According to [18], speed is the inverse of the drifting time in the sense that a higher speed is related to a lower number of time steps and a lower speed is related to a higher number of time steps. Therefore, the drift speed V_d is calculated by

$$V_d = \frac{1}{t_{de} - t_{ds}} \tag{2.2}$$

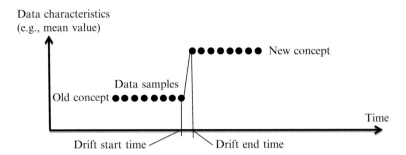

Fig. 2.11 Abrupt drift

where t_{de} and t_{ds} are, respectively, the time when the drift ends and the time when the drift starts.

According to its speed, a drift can be either abrupt or gradual:

- Abrupt drift occurs when the new concept suddenly replaces the old one in short drifting time. This kind of drift immediately deteriorates the learner performance, as the new concept quickly substitutes the old one (see Fig. 2.11).
- Gradual drift occurs when the drifting time is relatively large. This kind of drift is hard to detect since it creates a period of uncertainty due to the cohabitation of both old and new concepts. Gradual drift can be either probabilistic or continuous:

  – Gradual probabilistic drift refers to a period when both new concept generated by the source S_2 and old concept generated by the source S_1 cohabit. In other words, there is a weighted combination between data samples generated by S_1 (old concept) and the ones generated by S_2 (new concept). As time passes, the probability of sampling from S_1 decreases, whereas the probability of sampling from S_2 increases until the new concept totally replaces the old one (see Fig. 2.12).
  – Gradual continuous or incremental drift corresponds to the case where the concept itself continuously changes from the old to the new concept, by suffering small modifications at every time step. Therefore, during the continuous or incremental change, the new concept does not yet appear; the patterns representing the continuous drift do not have the same characteristics. It is worth to mention that these changes are so small that they are only noticed during a long time period (see Fig. 2.13). When the continuous drift is ended, then the new concept appears, and starting from this time, the incoming patterns are generated from the same source.

Example 2.7: Probabilistic Gradual Drift Let us suppose that the incoming data samples arrive within batches. Each batch contains 100 data samples. Let us suppose that the patterns or data samples in the first batch are generated by the source S_1. Then let us suppose that in the second batch, 80 patterns were generated by the source S_1 (old concept), while 20 patterns were generated by the new source S_2 (new concept). In the

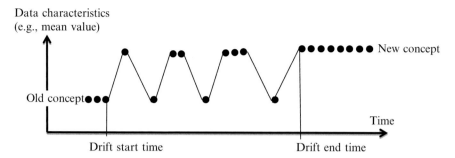

Fig. 2.12   Gradual probabilistic drift

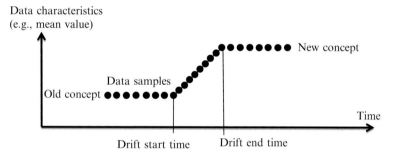

Fig. 2.13   Gradual continuous drift

third batch, 50 patterns are generated by the S1, and the other 50 patterns are generated by the new source S_2. Finally, in the fourth and fifth batches, all the patterns are generated by the new source S_2. Therefore, this drift replaces gradually the old concept (generated by S_1) by the new one (generated by S_2).

Figure 2.14 shows the probabilities of occurrence of patterns belonging to the old and new concepts. We can see that the probability P_1 of occurrence of patterns generated by S_1 decreases, while the one P_2 for patterns generated by S_2 increases over time. At the end, P_1 will be equal to 0, while P_2 is equal to 1 in order to indicate that the new concept has replaced completely the old one.

2.4.2   Drift Severity

The severity refers to the amount of changes caused by the drift occurrence. The drift severity can be high or low (partial). A high or global severity (see Fig. 2.15b) means that the old concept has been completely changed. Therefore, the whole region occupied by this old concept will be impacted by the drift. A low or partial severity (see Fig. 2.15c) refers to a change impacting only a part of the region

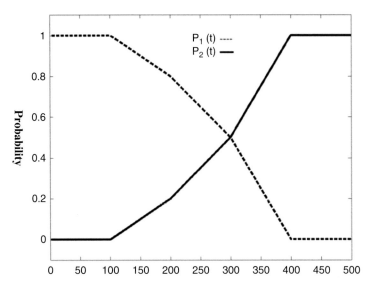

Fig. 2.14 Probability P_1 of occurrence of patterns from old concept and the probability P_2 of occurrence of patterns from new concept for the case of probabilistic gradual drift

occupied by the old concept in the feature space. Therefore, patterns belonging to both old and new concepts will cohabite.

Let us take the user preferences for document retrieval. He may change completely his search criteria for documents. Therefore, the new documents do not share any similarity with the old ones. In other words, no old document belongs to his preferences anymore. He may also change only some of his criteria for document search. In this case, some old documents remain belonging to his preferences.

2.4.3   Drift Influence Zones

Concept drift can be global (see Fig. 2.15b) or local (see Fig. 2.15c) according to the impacted zone of the feature space by the drift.

Local concept drift is defined as changes that occur in some regions of the feature space. Hence, the time required to detect the local drifts can be arbitrarily long. This is due to the rarity of date samples belonging to the new concept since both old and new concepts cohabite. Moreover due to this cohabitation between old and new concepts, data samples generated from the new concept can be considered as noises, which makes the model unstable. Hence, to overcome the instability, the model has to (1) effectively differentiate between local changes and noises and (2) deal with the scarcity of data samples that represent the local

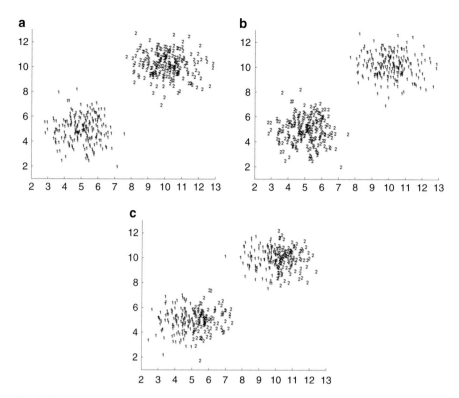

Fig. 2.15 Difference between drifts of partial and global severity. (**a**) Initial two classes W_1 and W_2. (**b**) Global drift impacting all the regions of both classes. (**c**) Local drift impacting local zones of the regions occupied by both classes

drift in order to effectively update the learner. The global concept drift is easier to detect since it affects the overall feature space. In such case, the difference between the old and the new concept is more noticeable, and the drift can be earlier detected. This is due to the fact that the old concept will not cohabite anymore with the new concept.

Example 2.8: Local Concept Drift Let us take the example of Fig. 2.16. Figure 2.16.a shows the decision boundary of the initial classifier built using the data samples of the batch B_1. In the second batch B_2, a local drift occurred impacting a partial zone of the feature space between the two classes (see Fig. 2.16b). Therefore, only the patterns located in this zone will be impacted by the drift, while the other patterns keep their initial classes labels. The decision boundary will be updated by taking into account only the patterns impacted by this local drift (see Fig. 2.16b). In the batch B_3, another local drift occurred. A new update of the decision boundary is required in order to maintain the classifier performance as we can see in Fig. 2.16c.

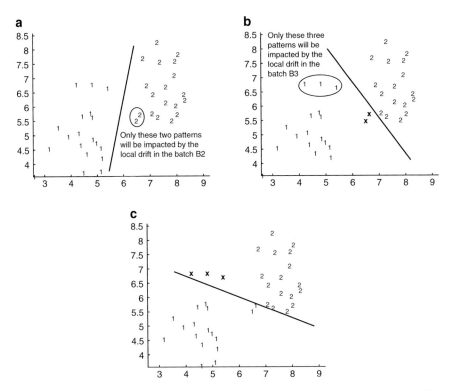

Fig. 2.16 Local drift impacting the feature space. The decision boundary of the initial classifier built using the data samples in the batch B_1 (**a**), after a first local drift in the batch B_2 (**b**) and a second local drift in the batch B_3 (**c**)

If we consider that class W_1 in Fig. 2.16b represents the viruses which are not resistant against antibodies while W_2 represents the viruses resistant to these antibodies, then some viruses of W_1 may develop a resistance against antibodies and become resistant while the other viruses remain as before not resistant. Therefore, this drift is just local drift concept.

2.4.4   Drift Occurrence Frequency and Recurrence

A concept may suffer from several drifts over time. If these drifts occur within regular time intervals, then they are called periodic drifts. Their occurrence frequency can be measured as the inverse of the number of time steps between two successive starts of drift. If these drifts occur in random or irregular time intervals, then they are called aperiodic drifts. If a concept suffers from the same drift at different time instances, then they are called recurrent drifts. In this case, concepts

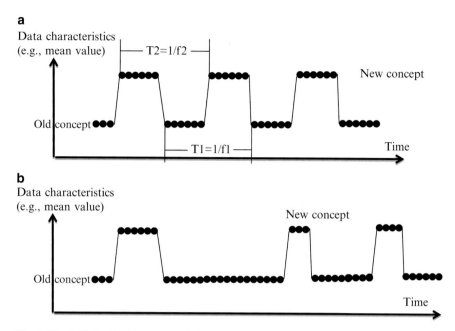

Fig. 2.17 Cyclic (periodic) recurrent drift (**a**) and acyclic (aperiodic) recurrent drift (**b**). f_1 is the occurrence frequency of the old concept, while f_2 is the occurrence frequency of the new concept

previously active may reappear after some time. Recurrent drifts may have *cyclic* or *acyclic* behavior:

- Cyclic recurrent drift (see Fig. 2.17a) may occur according to a certain periodicity or due to a seasonable trend. For instance, in electricity market, the prices may increase in winter due to the increase of demand and then return to previous level in the others seasons. The weather prevision is another example for cyclic recurrent drift where the prevision rules change in cyclic manner according to the active season.
- Acyclic recurrent drift (see Fig. 2.17b) occurs in aperiodic or random time intervals. For instance, the electricity prices may suddenly increase due to the increase of petrol prices (because of a political or economic crisis) and then return to previous level when petrol prices decrease.

It is worth underlining that when an old concept reappears, it may not be completely similar to its initial case. As an example, let us take a machine with two classes representing the normal and failure operation conditions. In the initial case, there is only the normal class. When the machine fails, a new concept occurs which is the failure class, and the old concept disappears which is the normal class. When the machine is repaired, it returns to the normal class (old concept). However, according to the efficiency of the maintenance actions and to the machine degradation status, the machine may not return completely to the class of normal operation conditions.

2.4.5 Drift Predictability

The predictability criterion was initially used in [18]. It indicates whether the drift is completely random or follows a pattern. Therefore, a drift is predictable if it follows a certain mechanism or set of rules. An example of predictable drifts is the weather forecasting. The change in the prediction rules is predictable according to the change in seasons.

A drift is unpredictable when its occurrence is random which does not follow any mechanism or rule. Example of unpredictable drifts is the occurrence of faults. Indeed, it is impossible to predict the occurrence of a fault before its occurrence. It can occur at any time and in different contexts.

It is interesting to consider the predictability of a drift for two reasons. First, when a drift is predictable, it is easier to understand its origins and expect its future effects (achieve a prognostic function). Second, a predictable drift can be accurately handled with a minimum delay of detection and false alarm rate, which is a desirable property in many real-world applications.

2.5 Drift Concept in Real-World Applications

There are multiple application domains in which concept drift plays an essential role. For these applications, the machine learning and data mining methods used to build a model (learner, classifier, etc.) for prediction or classification must take into account the concept drift in order to maintain the model performance and accuracy. However, in real-life applications, the concept drift may be complex or diverse in the sense that it presents time-varying characteristics. As an example, a concept drift can be recurrent, gradual, and local at some time instances, and then it becomes abrupt and global at other time instances. Hence in reality, often mixtures of many characteristics of drift can be observed during the transition phase of concept drift [21].

In [19, 21], the applications, where concept drift problem is relevant, are classified into four families:

- Monitoring against adversary actions as intrusion detection [22] or for management as traffic management and control [23]
- Personal assistance and information as recommender systems [24], customer profiling [25], and spam filtering [26]
- Predictions for decision making as evaluation of creditworthiness [27], electricity prices prediction [3], and sales prediction [28]
- Ubiquitous environment applications including a wide spectrum of moving stationary systems which interact with changing environments as moving robots [29] and smart house appliances [30]

In order to design an efficient self-adaptive model for an application where concept drift is related, the following points must be determined:

- Objective of the model: classification or regression
- Sources of the drift: environmental, system itself, or both
- Characteristics of the drift: speed, severity, predictability, etc.
- Speed of learning and required data loads
- Required accuracy and costs of mistakes
- Availability of true labels and kind of feedback
- Data samples balanced or imbalanced

As we will see in Chap. 3, these points allow to understand the drift phenomenon of the data streams generated by the application and therefore to guide the choice toward the suitable methods and tools to design an efficient self-adaptive learning scheme.

Example 2.9: Characterizing the Concept Drift Related to the Problem of Fault Diagnosis of a System Fault diagnosis aims at deciding at each instant whether a system works in normal operation conditions or a failure has occurred. The occurrence of a fault entails a drift in the system's normal operation conditions. This application has the following aspects:

- *Objective of the model* is the classification. The model is a classifier able to assign a new pattern, representing the current operation conditions, to one of two classes: normal or failure.
- *Sources of the drift* are exogenous caused by the system environments as a cut of physical connection between two switches because of an external action or the degradation of a service quality over time due to its wearing or the accumulated pollution, etc.
- *Characteristics of the drift*:

    - Two types of faults may occur: permanent and intermittent faults. Permanent fault can be either abrupt or gradual. If it is abrupt, then the drift is a shift in the system operation conditions from normal to faulty. While if it is gradual, then the drift is a degradation in the system performance. In this case, it is a continuous or incremental drift. If the fault is intermittent, then the drift is gradual probabilistic since patterns representing both normal and faulty classes cohabite. Then the number of fault patterns increases over time until the failure takes over completely.
    - The drift speed in both cases (gradual probabilistic and gradual continuous) depends on external factors generating these degradations, as the rate of pollution, moisture, temperature, etc.
    - The drift is acyclic recurrent since the fault may occur at any time and may be eliminated thanks to the maintenance actions.
    - The fault may be local (low severity) and/or global (high severity) impacting partially or completely the feature or instance space. As an example, a small degradation will generate a local drift since the system keeps an acceptable

performance which is not far from the nominal one. Example of local drift is the case where a pump is partially failed off or failed on. While an abrupt and severe fault generates a global drift as the case for a pump failed off or failed on completely.

- *Speed of learning* must be fast since the decision about the system status (normal/faulty) must be taken online in order to limit the fault adversary impacts.
- *Required accuracy and costs of mistakes* depend on the application criticism as well as the impact of faults on the system performance. For instance, if the system is a nuclear reactor, then the required accuracy of detecting faults is very high.
- *Availability of true labels (normal/fault)* is delayed. The true labels whether the system was in normal or failure operation conditions come available only after certain time (inspection, maintenance). These labels are hard; the system is either normal or faulty.
- *Data samples are highly imbalanced* since the data samples belonging to normal operation conditions are much more bigger than the ones representing a fault.

Chapter 3
Handling Concept Drift

3.1 Introduction

When the incoming data streams become nonstationary and they distribute on (source) changes over time, traditional online learning algorithms cannot provide an efficient model (e.g., classifier) adapted to this context. This is because they are unable to discard outdated data, i.e., unlearn old concepts. Hence, self-adaptive, or dynamic, learning algorithms were developed in order to:

- Continuously allow learning models to evolve in response to changes over time.
- Handle and react to changes in incoming data characteristics.
- Incorporate forgetting mechanisms and discard data from outdated concepts.

In increasing number of real-world applications, data are presented as streams that may evolve over time, and this is known as concept drift. Handling concept drift is becoming an attractive topic of research that concerns multidisciplinary domains. Therefore, multiple methods and approaches have been proposed in the literature to lean from evolving and nonstationary data streams. The performances of these methods and techniques depend on the application context and constraints. It is not clear in the literature how the choice of an appropriate method or technique can be made for a given application.

Consequently in this chapter, the different methods and techniques used to learn from data streams in evolving and nonstationary environments will be presented, and their performances will be compared according to the generated drift characteristics as well as to the application context and objectives. The goal is to define the criteria to be used in order to help readers to efficiently design the suitable learning scheme for a particular application. For this aim, these methods and techniques are classified and compared according to a set of meaningful criteria. Several examples will be used to illustrate and discuss the principal and the performance of these methods and techniques.

© The Author 2016
M. Sayed-Mouchaweh, *Learning from Data Streams in Dynamic Environments*,
SpringerBriefs in Applied Sciences and Technology,
DOI 10.1007/978-3-319-25667-2_3

3.2 General Learning Scheme to Handle Concept Drift

Three major tasks (blocks) must be included in a learning scheme in order to learn from data streams generated in evolving and nonstationary environments. These tasks are monitoring, updating, and diagnostics tasks.

The monitoring task allows dealing with the instability of the learner when drifts occur in the sense that they integrate the new information carried out by the drifting data samples and delete or unlearn the obsolete ones. There are two ways to achieve that according to how the machine learning methods switch on the learner adaptivity. They are either trigger based or evolving. In the trigger based, the learner is adapted only when a drift is detected. Therefore, a set of change detectors must be employed. On the contrary, the evolving methods do not detect changes or drifts, but they keep updating the learner at regular intervals in order to maintain its performance.

The updating task aims at reacting to the occurrence of drifts by updating the learner in order to preserve its performances. This update is achieved by using the data samples representing or describing the drifts. The learner update depends on how the drifts are monitored (monitoring task). In the case of trigger-based approaches, the learner is updated when drifts are detected, and the learner can either be relearned from scratch or updated using a recent selection of data (data window). In the evolving methods, the learner is updated as new data sample is available regardless a drift has occurred or not. This continuous update is achieved either by discarding the oldest data sample (decremental learning) and integrating the newest one (incremental) or according to the performance (feedback) of the learner regarding the decision of the new data sample.

The diagnostic task looks for interpreting the detected changes in concepts. This interpretation is then used as a short-term prognosis about the tendency of the future development of the current situation. This prognosis is useful to formulate a control action to modify the dynamics of a system. For instance, let us suppose that we have two classes A and B. Let us suppose that class A represents the normal state of a system, while class B is a fault state. When the diagnostic task provides the result "the system's state has been moved away from class A and is approaching class B," this means that the system needs to be repaired, adjusted, or reconfigured. The goal is to inverse its tendency, to move toward a fault state, by forcing it to return to the normal functioning state. In addition, this task may provide the remaining useful life (RUL) of a system before the failure. RUL is used in condition-based maintenance (CBM) to schedule required repair and maintenance actions prior to breakdown (failure state). Let us take the example of Fig. 3.1 showing a fault propagation case. If one catches the fault at 5 % severity, one needs to replace only the component. If the fault is not caught until 10 % severity, the subsystem must be replaced, and at failure, the entire system must be replaced [31]. Thus, predictions about fault severity and impending failures are very useful.

Figure 3.2 illustrates the position of these tasks required to convert a learning scheme into self-adaptive one.

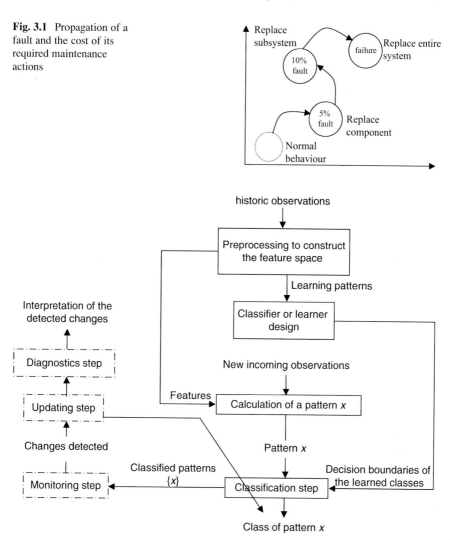

Fig. 3.1 Propagation of a fault and the cost of its required maintenance actions

Fig. 3.2 Self-adaptive learning scheme

3.3 General Classification of Methods to Handle Concept Drift

There are several machine learning and data mining approaches that are proposed to handle concept drift. These approaches can be classified according to how:

- Decision boundaries are built either by one learner or by a set of learners (ensemble learners working in competition or in collaboration).

- Data samples are processed to build decision boundaries and to monitor the drifts either sequentially (treating one data sample at a time) or in parallel (treating multiple data samples through a time window).
- Drifts are monitored either by using supervised change detectors (true labels of incoming patterns are available shortly after their arriving) or unsupervised change detectors (the true labels are unavailable or delayed).
- Concept drift is handled, and the decision boundaries (or model parameters and structure) are updated in response to the concept drift occurrence either explicitly using informed or change trigger methods or implicitly using blind or evolving methods.

Figure 3.3 resumes this general classification of methods and techniques of the literature used to handle concept drift. In the following sections, the methods and techniques classified according to these criteria will be detailed.

3.4   Informed Methods to Handle Concept Drift

Informed methods explicitly detect drifts using triggering or drift detection mechanisms. The latter may monitor the performance of a learner [20, 21, 32], the data distributions (i.e., a change in the data sample characteristics) [33–37], or the learner's structure and parameters [38] in order to detect a drift. When a drift is detected, the informed methods either relearn the model from scratch or update it using a recent selection of data samples through a time window or a register.

However, it is worth to underline that when the informed methods are used, preserving the learner performance is not the only concern but also controlling the false alarms and the missed detections. For instance, a false alarm that a fault has occurred in a machine as a generator may lead to stop the latter and therefore to reduce its production. Moreover, missing the occurrence of fault in the generator may lead to increase significantly the maintenance costs due to the fault propagation and development (see Fig. 3.1).

The informed methods are based on three steps: (1) detecting the drift using change or drift monitoring indicators, (2) deciding the data samples to be used to

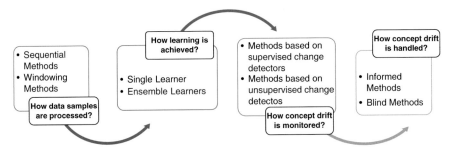

Fig. 3.3 General classification scheme of methods and techniques used to handle concept drift

update the learner (which data samples to keep and which ones to forget), and (3) deciding how to react or update the learner using the collected data samples.

The choice of method to handle drifting data samples depends on the availability of prediction feedback. If the true labels are immediately or shortly available after the prediction, then methods based on supervised drift or change indicators may be used. However, if data are partially labeled and the prediction feedback is delayed, then methods based on unsupervised drift or change indicators are the most appropriate. In the next, the tools and techniques used to achieve these three steps are detailed.

3.4.1 Methods Based on Supervised Drift Monitoring Indicators

Methods based on supervised indicators are useful for detecting changes when the prediction feedback is immediately or shortly available. These methods focus on preserving the learner performance by handling real concept drifts. The main key for handling this kind of drifts relies on monitoring the learner feedback indicators [20, 21, 39, 40] defined by 1.9, 1.10, 1.11, and 1.12:

- *Accuracy*: is calculated as the sum of correctly classified data samples divided by the total number of classified ones (see Example 1.9). Generally for handling drift, some approaches monitor the frequency of classification errors [39] or the distance (i.e., number of data samples) between two errors of classification [20, 40]. When this frequency or distance increases significantly beyond a certain threshold, this means that the classifier performance has been significantly decreased due to the occurrence of a drift in its environments.
- *Specificity*: is a measure of the ability of a learner to predict correctly data samples from a specific class, generally the positive class, according to all the data samples assigned by the learner to this specific class (see Example 1.10).
- *Recall or sensitivity*: is a measure of the ability of a learner to predict correctly data samples from a specific class, generally the negative class, according to all the data samples assigned by the learner to this specific class (see Example 1.11).
- *Precision*: is a measure of the ability of a learner to classify correctly the data samples from a specific class according to all the data samples belonging initially to this class (see Example 1.12).

Specificity and sensitivity are two measures that are commonly used in two class problems where one of the classes is severely underrepresented in the data set. Considering the example of fraud detection (Examples 1.3 and 1.9), the sensitivity concerns the class of fraudulent transactions, and it measures how many fraudulent transactions are detected, whereas the specificity concerns the class of genuine transactions, and it measures how many genuine transactions are identified.

The specificity and sensitivity are also suitable for monitoring a drift in the class prior probability that causes class imbalance.

These supervised indicators have the advantage of being reliable and independent of the used machine learning method. However, they operate in supervised mode, i.e., they require the availability of true class labels, which in turn can delay the detection of changes if the true label is not immediately or shortly available, as it is the case in most real applications.

Example 3.1: Informed Methods Based on the Use of Supervised Drift Monitoring Indicators to Handle Concept Drift Let us take the case of two Gaussian classes described in two-dimensional space as follows:

$$x \in W_1 \Rightarrow x \sim N(M_1 = (2, 1), \Sigma_1 = (1, 1))$$
$$x \in W_2 \Rightarrow x \sim N(M_2 = (10, 9), \Sigma_2 = (1, 1))$$

Let us suppose that an abrupt drift has occurred in W_1 leading to a change in its mean values as follows:

$$x \in W_1 \Rightarrow x \sim N(M_1 = (3, 12), \Sigma_1 = (1, 1))$$

This change in the parameters (mean values) of the normal law generating the data samples is represented as a shift of the location of W_1 in the feature space (see Fig. 3.4). This shift changes the data spatial distribution of W_1 in the feature space and impacts the classifier's decision boundary. W_1 before the shift will disappear (deleted). The incoming patterns from the drifted class W_1 will be misclassified by the classifier since the latter will assign them to W_2 while they belong to W_1 (see Fig. 3.4). If the drift indicator is the misclassification rate (error of prediction defined by Example 1.9), then the value of this indicator will increase over time. When the value of this indicator is greater than a threshold (see Fig. 3.5), then the decision boundary of the classifier must be updated using the recent patterns representing the new drifted concept (new W_1).

Fig. 3.4 Abrupt drift in the class W_1 leading to a shift in the feature space. This shift in W_1 impacts the decision boundary of the classifier

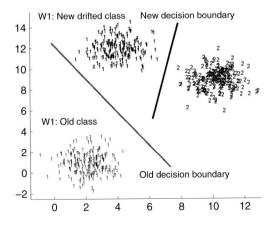

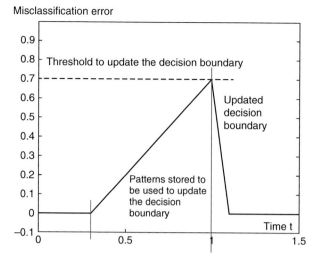

Fig. 3.5 Misclassification error before and after the decision boundary update of the classifier in response to the drift occurrence

3.4.2   Methods Based on Unsupervised Drift Monitoring Indicators

Methods based on unsupervised indicators are useful for detecting changes when the prediction feedback is unavailable or delayed. This is interesting for many real-world applications where data are unlabeled or partially labeled. Moreover, they can be useful for handling virtual concept drift since the latter does not impact the decision boundaries.

These indicators are based on the observation of dissimilarity in the data characteristics or distribution in the feature space as well as in the model complexity. The dissimilarity can result in time, in space, in both, or in the learner complexity in response to the occurrence of a drift as follows:

- Dissimilarity in time which results by a change of data distributions in the classes over time. In a specific zone or area in the feature space, the data distributions in the different classes will evolve over time [41]. Generally in the literature, the dissimilarity in time can be observed using hypothesis tests like: sequential probability ratio test (SPRT) [42], cumulative sum (CUSUM) test [43], Page-Hinkley test [13], etc.
- Dissimilarity in space which describes how data distributions evolve in the feature space. This dissimilarity entails classes to move and change the zones that they occupy in the feature space. Generally in the literature, the dissimilarity in space can be quantified using distance measures or metrics as: Euclidean distance [44], heterogeneous Euclidean overlap distance [45, 46], Mahalanobis distance [47, 48], Hellinger distance [34, 49–52], entropy measure [53], etc.

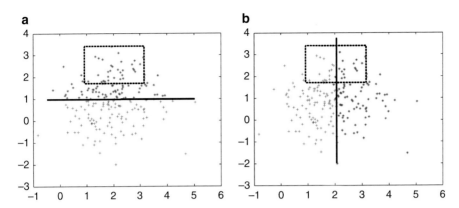

Fig. 3.6 Dissimilarity in time of data sample distribution into two classes (Star and Plus classes): (**a**) data generated from distribution D_1, (**b**) data generated from distribution D_2. Data samples in the *rectangle* are similar in space, but their class label is variable in time

- Model complexity which is related to the structure and/or the parameters of the model. For instance, the explosion of the number of rules for rule-based classifiers or the number of support vectors for support vector machine (SVM) method can give an indication about an unusual model behavior [54]. This type of indicators can perfectly operate in unsupervised mode. However, they can only be applied to some specific classifiers or learner for which the model complexity can be quantified.

Example 3.2: Concept Drift as a Dissimilarity in Time Let us consider a rotating decision boundary between two classes (Star and Plus classes in Fig. 3.6). In the initial case, the data samples are generated from the data distribution D_1 (see Fig. 3.6a). The data samples in a specified area of the feature space (e.g., the rectangle in Fig. 3.6a) represent the data sample distribution in each of both classes at a certain time. For the example of Fig. 3.6a, these data samples belong only to the Star class. In Fig. 3.4b, the data samples are generated by the distribution D_2. The data sample distribution in the rectangle (specified zone in the feature space) changed since the data samples (Star points) belong now to both Star and Plus classes. Therefore, the data sample distribution in the feature space evolved over time.

Example 3.3: Concept Drift as a Dissimilarity in Space The example of Fig. 3.4 represents an abrupt drift (shift) in the class W_1 entailing a change in the zone occupied by W_1 in the feature space. The data samples of W_1 were generated by the distribution D_1. Then, the latter changed, and the new data sample distribution of W_1 is generated by the new distribution D_2. This change of data sample distributions is represented by a change in the zone of the feature space occupied by W_1. This change generates dissimilarity in space between the data samples of W_1 generated before and after the shift.

3.5   Drift Handling by Single Learner

In evolving and nonstationary environments, single self-adaptive learning methods are widely used for handling concept drift. In these methods, only one classifier or model is used to achieve the classification or the prediction. They can be considered as an extension of incremental learning algorithms as follows (see Fig. 3.7):

- They incorporate forgetting mechanisms in order to discard data from outdated concepts. This was initially known as decremental learning [54]. The idea behind decremental learning is to integrate forgetting capacity in order to unlearn the outdated data samples, or to reduce their weight, since they become obsolete due to environment changes.
- They integrate a self-adaptive mechanism in order not only to adapt the learner after the drift occurrence but also to describe this drift in order to provide maintenance instructions to recover or to accommodate the consequences of this drift. The development of such self-configuring or self-repairing systems is a major scientific and engineering challenge [55].

Using a single learner can be interesting for controlling the complexity of the system since updating one model can be achieved in real time with reduced computational efforts. However, approaches handling drifts using one learner are not recommended for handling recurrent drifts. This is due to the fact that as they process online, they are continuously adapted to the current concept. Hence, when a previous concept reoccurs, these approaches relearn it from scratch without getting benefit from its previously existence.

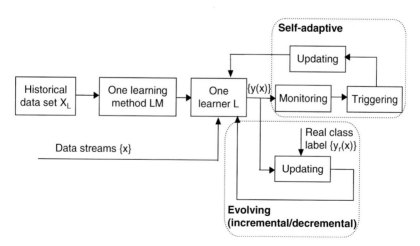

Fig. 3.7 Self-adaptive or evolving scheme to handle concept drift based on the use of a single learner

3.6   Drift Handling by Ensemble Learners

Ensemble learners [56] are based on the use of several classifiers, models, or learners built by one learning method with different configurations or by different learning methods in order to achieve the classification or the prediction of a new incoming pattern. The aim of using ensemble methods is to achieve more accurate prediction (e.g., classification) on training data set and better generalization on unseen data samples [57]. The approaches based on ensemble learners have received great interest since they have shown better generalization abilities than single learners. However, the strength of ensemble learners lies in the complementarity or diversity of its individual learners [58].

Therefore, the drawback of one learner can be avoided by the use of another learner which leads to obtain a global performance of the combination of learners at least greater than the one of each individual learner. Along this direction, ensemble learners appear to be promising approaches for tracking evolving data streams. They provide a natural way for adapting to changes either by modifying their structure (combination of the individual learners' decisions, selection of one individual learner's decision, etc.) or updating the combination rules used to fuse the learners' individual decisions into one global decision (classification, prediction).

The success of the ensemble methods for handling concept drift relies on two essential points: diversity and adaptability. The diversity refers to the complementarity of the individual classifiers. Therefore, the diversity allows ensemble learners to detect a complex drift with different characteristics. In addition, it allows the ensemble to recover faster from the drift by returning to the initial performance before the drift. The adaptability defines how ensemble learners will adapt to the concept drift. The adaptation can be achieved either by modifying the ensemble structure, retraining ensemble's individual learners, replacing old or obsolete individual learners with new ones, or updating fusion or combination decision rules (see Fig. 3.8).

Approaches based on ensemble learners handle concept drift using one of the following techniques (see Fig. 3.8):

- Training set management which indicates how data samples are managed for training and adapting the ensemble learners in response to concept drift
- Structure management which describes how individual learners are managed among the ensemble in order to react to the concept drift
- Individual learners' decision management which defines the rules to combine or fuse the individual learners' decisions and how these rules are updated in response to the drift occurrence

These ensemble's techniques to handle concept drift are detailed in the following sections.

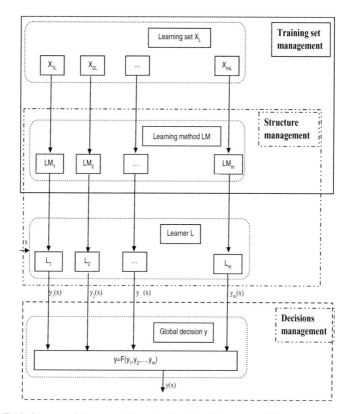

Fig. 3.8 Techniques used by ensemble methods to handle concept drift

3.6.1 Ensemble's Training Set Management

Training set management aims at partitioning efficiently the training data set in order to train the individual learners and to increase their diversity and adaptability in response to concept drift. To achieve that, the training set is partitioned using either block-based or weighted data techniques.

3.6.1.1 Block-Based Technique

In block-based technique, the training set is divided into blocks or chunks of data. Generally, these blocks are of equal size, and the learning, evaluation, or updating of individual learners is achieved when all data samples in a chunk are available. The individual learners are evaluated periodically, and the weakest one is replaced by a new (candidate) learner after each data block [56, 59–61]. This technique preserves the adaptability and the diversity of the ensemble's individual learners in such way that learners trained in recent blocks are assigned high weights since they are considered to be the most suitable for representing the current concept.

Two advantages for block-based ensembles can be cited. Firstly, they are suitable for handling continuous drift which is a challenging task since detecting this drift requires long period of time. Since the individual learners in the block-based ensemble are trained in different periods of time, they can react properly and efficiently to such gradual continuous drifts. Secondly, they are widely used in real-world applications where the true labels are not immediately provided. Therefore, the individual learners can be trained when the true labels are available. For instance, in credit card fraud detection, the true labels (i.e., authentic/fraud) of credit card transactions are usually available in the next billing cycle.

The main drawback of block-based ensembles is to tune in adaptive manner the block size in order to obtain a compromise between fast reactions to drifts and high accuracy. If the block size is too large, block-based ensembles may slowly react to abrupt drift, whereas small size can impact adversely the performance of the ensemble in time periods where the concepts are stable. Moreover, blocks of small size increase the computational costs.

3.6.1.2 Weighted Data Technique

In weighted data technique, the individual learners are trained according to weighted data samples from the training set. Popular instance weighted technique is used in the online bagging ensemble and online boosting ensemble [62]. Both takes a single learning method and generate ensembles.

In online bagging ensemble, M individual learners are trained by using k copies of each new data sample in order to update each individual learner or model. Therefore, a training data sample is presented k times where k is a weight drawn from a Poisson(1) distribution.

In online boosting [62], the process to weight data samples is slightly different from online bagging. The weight k of each data sample is drawn from a Poisson(λ) distribution, where λ is increased for the individual learner L_i when this data sample is misclassified by the previous learner L_{i+1} and decreased otherwise. Therefore, the weighting process in online boosting intensifies the reuse of misclassified data samples in order to increase the diversity and adaptability by training more of individual learners using the misclassified data samples.

Methods based on online bagging and online boosting [18, 20] can be useful for:

- Class imbalance where some classes are severely underrepresented in the data set
- Local drift where changes occur in only some regions of the instance or feature space

This is thanks to the weighting process which intensifies the reuse of underrepresented classes and helps to deal with scarcity of data samples that represent the local drift. However, the main drawback of these approaches is the increase of

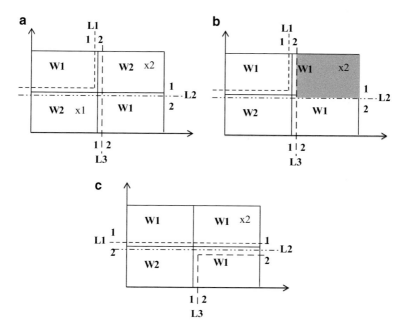

Fig. 3.9 Update of the individual learners in response to the occurrence of a drift based on the use of training set management

computational costs as data samples grow up. Indeed, since these methods are based on duplicating instances, they may lead to an expensive consumption of resources in time periods when the concepts are stable.

Example 3.4: Ensemble's Training Set Management to Handle Concept Drift Let us take the example of Fig. 3.9a showing two classes W_1 and W_2 in two-dimensional feature space. There are three learners (classifiers) L_1, L_2, and L_3 whose decision boundaries are depicted in Fig. 3.9a. The latter shows two patterns, x_1 and x_2, belonging to W_2. The classification decisions for x_1 according to the individual learners are: $L_1 : y(x_1) = 2; L_2 : y(x_1) = 2; L_3 : y(x_1) = 1$. According to the majority vote, the ensemble will issue the classification decision: $y(x_1) = 2$. Likewise for the classification of x_2: $L_1 : y(x_2) = 2; L_2 : y(x_2) = 1; L_3 : y(x_2) = 2$. Therefore, the ensemble decision will be $y(x_2) = 2$. Through the classification of these two patterns, we can notice the interest of the diversity of classifiers to obtain a correct classification decision for incoming patterns.

Let us now suppose that a drift occurred as we can see in Fig. 3.9b. The impacted zone by the drift in the feature space is marked by the gray zone in Fig. 3.9b. In this impacted zone, the patterns do not belong anymore to W_2. They belong now to W_1 due to the drift occurrence. Consequently, the data samples in the current data chunk, representing this drift, will be misclassified by L_1 and L_3, while L_2 will correctly classify these samples (see Fig. 3.9b). For instance, the pattern x_2 in Fig. 3.9a, which belongs to the impacted zone by the drift, was classified before the drift by the ensemble learners as follows: $L_1 : y(x_2) = 2; L_2 : y(x_2) = 1; L_3 : y(x_2) = 2 \Rightarrow y(x_2) = 2$. Hence, x_2 is misclassified by the ensemble after the occurrence of the drift. Consequently, the ensemble prediction accuracy will be decreased. When the impacted learners by the drift (L_1 and L_3) are trained using the recent data chunk, representing the drift, their decision boundaries will be updated as it is depicted in Fig. 3.9c. Thanks to this decision boundary update, the ensemble classifies correctly the patterns coming from the new concept, and therefore it maintains its performance after the drift occurrence. For instance, the classification decisions for x_2 according to the individual learners after the update of L_1 and L_3 are: $L_1 : y(x_2) = 1; L_2 : y(x_2) = 1; L_3 : y(x_2) = 1$. According to the majority vote, the ensemble will issue the classification decision: $y(x_2) = 1$.

3.6.2   Ensemble's Structure Management

Another way to ensure a good adaptability to drift is to efficiently manage the structure of the ensemble. Hence, the adaptability can be fulfilled either through a fixed or a variable ensemble size.

In fixed ensemble size, the size refers to the number of base or individual learners in the ensemble. In many approaches, the size is a priori fixed and generally chosen experimentally. One of the popular approaches for managing fixed-size ensemble is to periodically evaluate base learners and replace the weakest learner with a new one trained on recent data [56, 60, 63–65]. Likewise, other approaches make use of drift detection mechanism into the ensemble in order to replace the weakest learner only if a drift is signaled [20, 66, 67]. The fixed-size ensemble can be of interest when the objective is to control the ensemble complexity.

In variable ensemble size [68, 69], the number of individual learners is automatically adapted. For example, in [68], a novel ensemble classifier system using unlimited pool of classifiers is proposed. In contrast to the fixed-size ensemble approached, individual classifiers or learners are not removed from the pool. Therefore, knowledge of past contexts is preserved for future use. This is beneficial for recurrent drifts. Only selected learners from the pool can join the decision-making ensemble. The process of selecting classifiers and adjusting their weights is realized by an evolutionary-based optimization algorithm that aims to minimize the system misclassification rate. The variable size ensemble can be interesting in real-world applications with recurrent drifts where the objective is to take benefit from past knowledge.

3.6.3   Ensemble's Final Decision Management

Producing one final decision based on the management of the decisions issued by the individual learners can be achieved by one of the following techniques:

dynamic weighting, dynamic selection, or combining both dynamic weighting and selecting techniques.

3.6.3.1 Dynamic Weighting Technique

Dynamic weighting technique [56, 60, 65, 67, 70, 71] is based on the combination of the decisions issued by the ensemble's individual learners using weighted majority vote. The key point of this technique is that the weight attributed to an individual learner depends on its prediction accuracy achieved on recent data blocks. Since the occurrence of a drift impacts adversely the accuracy prediction of learners, then the weight updating of these learners over time allows to react to the drift and maintain the overall prediction accuracy of the ensemble. The learners are trained on previous data blocks, and the performance of each learner is evaluated by estimating its expected prediction error on data samples from the most recent data blocks. Thereby, the learner that achieves the highest accuracy is attributed the best weight. Consequently, the base learners' weights are updated over time according to the changes in their environment.

The major drawback of the ensemble approaches based on this technique is that the corresponding weights are computed according to previous data chunk that may present old concepts. However, when facing to new drifting data chunk, these weights may not reflect the new concept. Hence, learners with high weight for previous data chunk may not represent the learner efficiency according to these new drifting data samples.

A possible remedy to overcome this drawback is proposed in [67] where the base learners are continuously trained online using recent data blocks. Hence, they guarantee that the ensemble is containing learners that currently represent new concepts.

The dynamic weighting technique can handle efficiently very slow gradual probabilistic and continuous drifts where changes are so small that they are observed during a long time period. This is because the learners' weights are updated using all the data samples within a chunk. However, they still have problems responding to abrupt drifts and leveraging prior knowledge of recurring contexts [70]. This is due to the fact that the weight update is achieved after the reception of the last data sample in a chunk. Hence, the reaction to an abrupt drift is delayed to the end of the reception of all the data samples within a data chunk.

Example 3.5: Dynamic Weighting Technique to Handle Concept Drift Let us take the example of Fig. 3.10a showing two classes W_1 and W_2 in two-dimensional feature space. There are two learners (classifiers) L_1 and L_2 whose decision boundaries are depicted in Fig. 3.10a. The latter shows a pattern x belonging to W_1. The classification decisions for x according to the individual learners are: $L_1 : y(x) = 1; L_2 : y(x) = 2$. In order to classify correctly x, L_1 must have a weight greater than the one for L_2. Hence, according to the weighted majority vote, the ensemble will issue the classification decision: $y(x) = 1$.

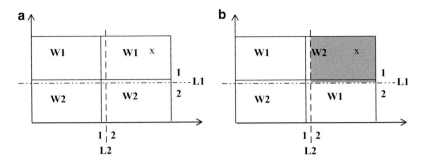

Fig. 3.10 Update of the individual learners' weights in response to the occurrence of a drift based on the use of weighted majority vote

Let us now suppose that a drift occurred as we can see in Fig. 3.10b. The impacted zone by the drift in the feature space is marked by the gray zone in Fig. 3.10b. In this impacted zone, the patterns do not belong anymore to W_1. They belong now to W_2 due to the drift occurrence. Therefore, the data samples in the current data chunk, representing this drift, will be misclassified by L_1, while L_2 will correctly classify these samples. For instance, the pattern x in Fig. 3.10a, which belongs to the impacted zone by the drift, was classified before the drift by the ensemble learners as follows: $L_1 : y(x) = 1; L_2 : y(x) = 2 \Rightarrow y(x) = 2$. Therefore, x is misclassified by the ensemble after the occurrence of the drift. In order to take into account the occurrence of this drift, the weight of L_1 will be decreased, while the weight of L_2 will be increased. Hence, the ensemble will classify correctly x after the drift occurrence using the weighted majority vote.

3.6.3.2 Dynamic Selection Technique

Dynamic selection technique is based on selecting one learner among the ensemble pool that is currently the most appropriate for making prediction. The selection procedure is based on choosing the learner that was trained on most recent data block with either the highest weight according to the prediction accuracy or the closest one to the data sample to be classified. For the latter case [72], each learner among the ensemble is trained according to a different subspace. Hence, each learner is responsible of the classification in a specific area of the feature space. Therefore, closeness of the learner to a data sample can be considered in terms of distance between the learner subspace and the data sample in the feature space.

The dynamic selection technique offers three interesting points:

- It represents a leveraged handling of recurring concepts where the objective is to take benefit from past knowledge. Hence, when a concept reoccurs, an existing learner can be selected rather than generating a new one.

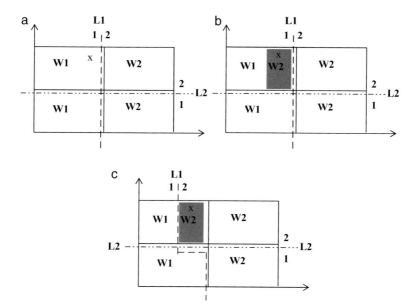

Fig. 3.11 Update of the decision boundary of the selected individual learner (L_1) in response to the occurrence of a local drift (*gray zone*)

- It is efficient for detecting local drifts where changes occur in only some regions of the feature space. In such situation, learner which is trained on this subspace is selected in order to provide accurate drift detection.
- It can help to control the complexity of the ensemble by selecting only the appropriate learners for decision making.

Example 3.6: Dynamic Selection Technique to Handle Local Concept Drift Let us take the example of Fig. 3.11a showing two classes W_1 and W_2 in two-dimensional feature space. There are two learners (classifiers) L_1 and L_2 whose decision boundaries are depicted in Fig. 3.11a. The latter shows also a pattern x belonging to W_1. The classification decisions for x according to the individual learners are: $L_1 : y(x) = 1; L_2 : y(x) = 2$. Since L_1 classifies correctly x, then the ensemble will select L_1, and the final classification decision will be $y(x) = 1$.

Let us now suppose that a local drift occurred as we can see in Fig. 3.11b. The impacted zone by the drift in the feature space is marked by the gray zone in Fig. 3.11b. In this impacted zone, the patterns do not belong anymore to W_1. They belong now to W_2 due to the drift occurrence. Since the drift is local, in the current data chunk, only some data samples represent the drift (the ones belonging to the impacted zone by the drift in the feature space).

Let us take the pattern x located in the impacted zone by the drift (see Fig. 3.11b). The classification decisions of x by L_1 and L_2 after the drift are: $L_1 : y(x) = 1; L_2 : y(x) = 2$. This means that L_1 is sensitive to the local drift since it misclassified x. Hence, the decision boundary of L_1 is impacted by this local drift.

While the decision boundary of L_2 is not impacted by this drift since L_2 classifies correctly x. Therefore, the ensemble will select L_1 and will update its decision boundary using the recent data chunk (see Fig. 3.11c). The classification decisions of x by L_1 and L_2 after this update are: $L_1 : y(x) = 2; L_2 : y(x) = 2$. This update of the selected learner (L_1) allows maintaining the ensemble prediction accuracy after the drift.

3.6.3.3 Combined Dynamic Weighting and Selection Technique

Combined dynamic weighting and selection techniques are based on selecting a subset of learners and then combining their predictions according to their weights. Generally, learners can be selected according to their prediction accuracy, age, or closeness to incoming data samples. Then, the final decision is calculated as the weighted aggregation of the decisions of these selected learners.

The approach developed in [68] is an example of this technique. In this approach, the classifiers are selected among unlimited pool of classifiers. The pool is unlimited because classifiers are not removed. Therefore, knowledge of past concepts is preserved for future use. This allows obtaining a leveraged handling of recurrent concepts.

The key point behind combining weighting and selection techniques is to get benefit from the advantages of the two techniques as follows:

- It allows handling both global and local drifts. In one side, thanks to the dynamic weighting process, the learners' weights are updated over time, and thus global drift can be efficiently handled. In the other side, the local drift can be accurately handled by selecting learners that are the most sensitive to these local changes in the feature space.
- It achieves a leveraged handling of recurring concepts thanks to the selection of a set of learners that have successfully handled similar concepts. Then, their decisions are combined according to their weights. Therefore, knowledge from previous concepts is preserved and appropriately reused.

3.7 Drift Handling by Sequential Data Processing

Given a sequence of independent random observations $X_1, \ldots, X_n$, where each X_i is generated from a distribution D_i, $i = 1, \ldots, n$, the online statistical analysis can be faced to one of these two situations:

- If all the observations $X_1, \ldots, X_n$ are generated according to the same distribution D_0, then we say that the distribution is stationary.
- If, for a subsequence $X_1, \ldots, X_k$ where $1 < k < n$, there exists a change point k $< \lambda < n$ such that X_λ is generated according to another distribution D_1, where

$D_0 \neq D_1$, then we say that the distribution is nonstationary; and therefore, D_0 and D_1 are, respectively, called pre- and post-change distributions.

The main objective of the sequential approaches is to check if there is a change between two distributions D_0 and D_1 at point λ. For this purpose, two hypotheses are posed:

- The null hypothesis H_0: $D_0 = D_1$ which is satisfied if there is no change in the original distribution D_0.
- The alternative hypothesis H_1: $D_0 \neq D_1$ which is satisfied if there is a change in D_0.

Hence, two parameters are defined in order to define the reliability level of the hypothesis testing:

- False alarm rate $\alpha = P(H_1|H_0)$ defining the probability of accepting H_1 when H_0 is true, i.e., the probability of detecting a change when the distribution is stationary
- Missed detection rate $\beta = P(H_0|H_1)$ defining the probability of accepting H_0 when H_1 is true, i.e., the probability of considering that the distribution D_1 is stationary ($D_1 = D_0$) while it is nonstationary ($D_1 \neq D_0$)

A change occurs when there is a significant difference between D_0 and D_1. This change is quantified using a dissimilarity measure $Diss_\lambda(D_0, D_1)$ allowing measuring the difference between the two distributions D_0 and D_1. Hence, the hypothesis testing compares $Diss_\lambda(D_0, D_1)$ against a change threshold τ in order to indicate whether the difference is significant or not. Hence, the dissimilarity measure and change threshold represent the two main parameters of hypothesis testing.
$Diss_\lambda(D_0, D_1)$ is determined by one of the following two techniques:

- Directly on data samples. For instance, $Diss_\lambda(D_0, D_1)$ can measure the dissimilarity between an incoming instance and a set of data samples. For this aim, some distance metrics can be used as: Euclidian distance [74], heterogeneous Euclidean overlap distance [45], Mahalanobis distance [47], and Hellinger distance [34].
- Based on summarized statistics from the two distributions like mean, variance, covariance, etc. For example, in [75], the exponentially weighted moving average chart was used for detecting a significant increase in the mean of the original distribution.

Change threshold τ can be either fixed or variable:

- The fixed change threshold [70, 76] is generally predefined by the user according to the specificity of the drift. A large threshold is suitable for gradual drift, whereas small threshold is suitable for abrupt changes. The fixed threshold can work well if the drift characteristics are a priori known; but this is rarely the case.
- The variable threshold is more suitable for handling different types of unknown drifts. In general, the sequential approaches which use variable threshold are

Table 3.1 Parameters of the hypothesis testing used by sequential approaches for drift detection

$X_1,..,X_n$	Sequence of independent random observations
D_0	Pre-change or initial distribution
D_1	Post-change distribution
H_0	Null hypothesis, i.e., there is no change in the original distribution D_0
H_1	Alternative hypothesis, i.e., there is a change in D_0
$\alpha = P(H_1|H_0)$	Probability of false alarm
$\beta = P(H_0|H_1)$	Missed detection rate
λ	Change point
$Diss_\lambda(D_0,D_1)$	Dissimilarity measure between D_0 and D_1 at point λ
τ	Change threshold

more autonomous and can effectively detect the change point [77, 78]. However, they have to control the false alarm and the missed detection rates.

Table 3.1 summarizes the parameters of the hypothesis test used by sequential approaches to detect changes (drifts).

3.8 Window-Based Processing to Handle Concept Drift

The learning methods based on the use of a window to handle concept drift process a set of data samples in order to detect a drift and to update the learner. A window is a short memory data structure which can store a set of useful data samples or summarizes some statistics concerning the model behavior or the data distribution in order to characterize the current concept. The window can be either data based or time based. In data-based window, also known as sequence-based window [79], the window size is characterized by the number of instances or data samples, whereas in time-based window, also known as timestamp-based window [79], the window size is defined by a duration or period of time.

The window size in both cases can be fixed or variable. In fixed window size-based approaches [80, 81], the window size is a priori fixed. However, the drift characteristics must be a priori known in order to handle efficiently the drift. Indeed for abrupt or fast speed drifts, small-size window is suitable for detecting quickly this drift, whereas large window size is suitable for detecting gradual or slow drifts. Variable size windows [82] have an adjustable size according to the drift characteristics, in particular its speed. Many techniques exist to determine dynamically the size of the window according to the drift characteristics as statistical hypothesis testing [20, 82] and control chart with variable change threshold [42].

The data samples within a window can be processed to update a specific learner. In this case, the data samples in the window are used to estimate the core statistic of the learner and maintain them consistent with the current concepts. For instance, the

developed approach in [83] (concept-adapting very fast decision trees) incrementally builds a decision tree from a data stream without the need for storing all the
instances used for training. Then, it uses a fixed data-based window in order to
continuously monitor the quality of previous split attribute. In this window, the core
statistic of each node in the tree is estimated and maintained consistent with the
current concept.

The data samples within a window can also be processed in order to update any
learner. In this case, the window is generic, and therefore, it uses some drift
monitoring indicators or change detectors based on the prediction performance
[20, 21] or on the characteristics of incoming data distribution [34, 36, 49, 53].

The window-based approaches to handle drifts can be also classified according
to how the windows evolve during the drift tracking (see Fig. 3.12) into single-
window-, two-window-, and multiple-window-based processing.

In single-window-based processing, the learner is periodically updated
according to the data samples stored in one window. The window may contain a
fixed number of data samples stored in the first-in-first-out (FIFO) data structure.
This means that when a new data sample arrives, it is saved in memory, and the
oldest one is discarded [79, 83]. The window may be also based on landmark in
order to store data samples starting from a given time point (timestamp) until a
certain condition is reached. For example, the landmark windows in [40] keep

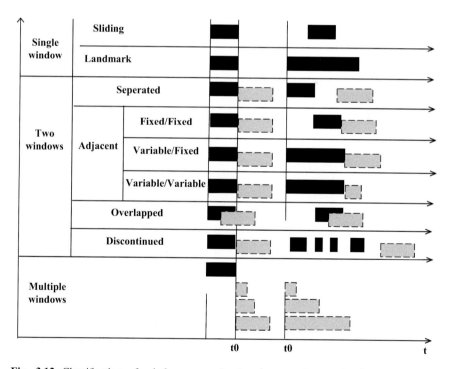

Fig. 3.12 Classification of window processing-based approaches to handle concept drift
according to how the windows evolve during the drift tracking

storing data samples instances and increase their size when there is no change. Once a drift is detected, the window size is reduced by containing the most representative data samples.

In two-window-based processing, two windows are used. The first window is used as reference to represent the initial concept and the second one to represent the current concept. These two windows can be separated, adjacent, overlapped, and discontinued:

- Separated where the reference and current windows diverge as data is processed (see Fig. 3.12). This strategy can be of interest for revising offline learners. Generally, the offline learner is initially trained on the reference window and then continuously tested through the current batch of data [84]. If the performance has decreased below a certain threshold, then the learner is relearned from the current data batch. It is worth to underline that separated windows are useful for handling gradual continuous drift. As it was discussed before, this kind of drift is hard to detect because changes are so small that they are only observed during a long time period. Therefore, when using separated windows, the dissimilarity between the reference and current windows increases as the two windows diverge, and the gradual continuous drift becomes easier to detect.
- Adjacent where the reference and current windows are kept related during incoming data samples processing. Generally, the comparison between reference and current windows is sequentially processed. Hence, an online learner is required. This online processing can be achieved by three variants according to the resizing process of both windows:
  - Fixed/fixed: The two windows are of fixed size [81]. This strategy is useful for handling abrupt drift but less efficient when the drift is gradual.
  - Variable/fixed: The size of reference window is variable, whereas the size of current window is fixed. The basic heuristic for resizing is to enlarge the window when no change is detected and shrunk it otherwise. This strategy is more useful for handling gradual drift [85].
  - Variable/variable: The size of both reference and current windows is variable. This strategy is used in [82], where the idea is to find the cut point that maximizes the dissimilarity between the two adjacent windows. Hence, along the data process, the size of the two windows is variable.

- Overlapped where the reference and current windows have data samples in common [74, 81] (see Fig. 3.12). This technique can be of interest with limited sampled data size or when the data set is imbalanced. The overlapped windows are generally used in the ensemble learners for accurately training individual learners. Thanks to this technique, the use of data is intensified in each learner in order to overcome the data scarcity.
- Discontinued where only subsets of the reference window are used for comparison with the current window [80, 81] (see Fig. 3.13). This technique selects subsets of data samples from the reference window according to some criteria. Then, it compares them to the current window. The subsets can be selected either

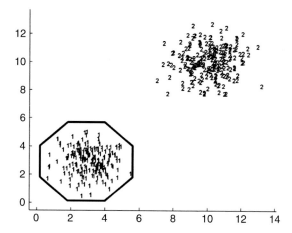

Fig. 3.13 Two classes into two-dimensional feature space with the decision boundary of class W_1

according to similarity in space, in time, or in data samples' representativeness of the encountered drift. The discontinued window technique is useful where the drift affects a little amount of data samples in the window (local drift) or for complex drift with different characteristics. Thanks to this technique, the comparison is more accurate, and the drift detection is more reliable.

Multiple windows were also used in the literature to handle concept drift and adapt dynamically to its characteristics (e.g., its speed). In [86], a set of three windows of variable size (small, medium, and large) was used in order to handle different speeds of drift. Each window is used to detect a specific drift speed: the small one is used to deal with very fast changing concepts, the middle is for slower changing concepts, and the large is to deal with very slow changing concepts.

3.9   Blind Methods to Handle Concept Drift

The blind methods implicitly adapt the learner to the current concept at regular time intervals without any drift detection. They discard old concepts at a constant speed independently of whether changes occurred or not. These approaches can be of interest for handling gradual continuous drifts where the dissimilarity between consecutive data sources is not quite relevant to trigger a change. This is due to the fact that during the continuous drift, the data samples are generated from different sources (see Sect. 2.4.1 for more details).

In order to handle concept drift implicitly, blind methods use one of the following techniques:

- Fixed-size sliding window [76], where the learner is periodically updated according to a fixed number of data samples stored in the first-in-first-out

(FIFO) data structure. Therefore, whenever a new instance arrives, it is saved in memory, and the oldest one is discarded.

- Instance weighting, where the learner is periodically updated according to the weighted instances from the training set. The instances can be weighted according to their age, i.e., the most recent data should have the highest weights [87] or according to their representativeness to the current concept using as example entropy measure [53].
- Ensemble learners, where the ensemble continuously adapts its structure to represent the current concept (see Sect. 3.6). A possible strategy is "replace the loser" where the individual learners are reevaluated and the worst one is replaced by a new one trained on recent data block.

The main limitation of blind methods is their slow reaction to drifts, because the updating process is the same whatever the drift is abrupt or gradual. Moreover, the regular updates can be too costly as the amount of arriving data samples may be overwhelming. These approaches are efficient only if the speed and severity of the change are known before or if rigorous instructions provided by an expert about the nature of the drift are available; but this is rarely the case.

Example 3.7: Blind Method to Handle Concept Drift Let us take the example of Fig. 3.13 showing two Gaussian classes in two-dimensional feature space. Figure 3.14 shows also the decision boundary representing the zone of class W_1 in the feature space. All the patterns located inside this decision boundary will be assigned to W_1. Let us suppose that at the initial time, the data samples of W_1 and W_2 are generated by the following Gaussian distributions:

$$x \in W_1 \Rightarrow x \sim N(M_1 = (3, 3), \Sigma_1 = (1, 1))$$
$$x \in W_2 \Rightarrow x \sim N(M_2 = (10, 10), \Sigma_2 = (1, 1))$$

Let us suppose that the distribution of W_1 suffered from an abrupt drift. The data samples are generated after this drift by the Gaussian distribution:

$$x \in W_1 \Rightarrow x \sim N(M_1 = (4, 4), \Sigma_1 = (1, 1))$$

This abrupt drift will entail a change in the zone of the feature space occupied by W_1 as we can see in Fig. 3.14.

In blind methods, no drift detection is used. A sliding fixed window is used to update the classifier. When the first pattern x of W_1 after the drift arrives, a blind method discards the first (oldest pattern) in the window and integrates the new arrived pattern (see Fig. 3.15). Then, the classifier parameters are updated based on the use of this new integrated pattern. In our example, the classifier parameters are the mean (M_1) and variance-covariance matrix (Σ_1). Both can be incrementally updated [88]. The sliding window will continue to integrate the new patterns generated by the new Gaussian distribution of W_1 and discarding the old ones.

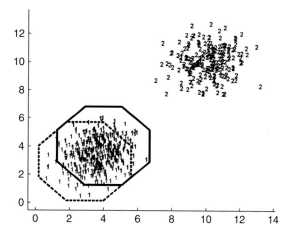

Fig. 3.14 Abrupt drift in the Gaussian distribution generating the data samples belonging to W_1. This drift entails a change in the zone of the feature space occupied by W_1

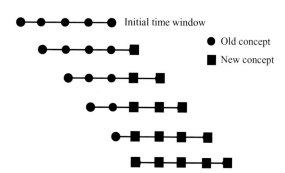

Fig. 3.15 Sliding time window used by a blind method to update the classes' decision boundaries

Then the classifier parameters are updated incrementally. This will lead, after including in the sliding window sufficient patterns from the new Gaussian distribution, to obtain the decision boundary of W_1 depicted in Fig. 3.15.

If we suppose now that another abrupt drift occurred in W_1 leading to generate data samples from the following Gaussian distribution (see Fig. 3.16),

$$x \in W_1 \Rightarrow x \sim N(M_1 = (4, 4), \Sigma_1 = (1, 1))$$

Again, the sliding window allows a blind method to update the classifier parameters (M_1 and Σ_1) in order to obtain the decision boundary depicted in Fig. 3.16.

Hence, the blind methods can react to the occurrence of drifts without the use of a drift monitoring indicator. However, they require a relative long time to react to abrupt drifts since they need a sufficient number of patterns generated by the new distribution in the sliding window in order to update the classifier parameters.

Fig. 3.16 Second abrupt
drift in the Gaussian
distribution generating the
data samples belonging to
W_1. This drift entails a new
change in the zone of the
feature space occupied by
W_1

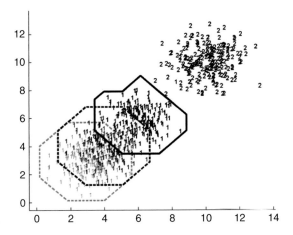

3.10 Drift Handling Evaluation

It is essential to evaluate the performance and the reliability of the designed
adaptive machine learning and data mining framework to handle drifts. This can
be done by defining some meaningful evaluation criteria. According to the requirements of the real-world applications, the evaluation criteria are related to autonomy,
reliability, parameter setting, and complexity of the designed scheme to handle
drifts.

Autonomy refers to the level of human involving in handling concept drift.
Generally, autonomous approaches can rely on either variable change threshold
[77, 78] or self-adaptive windowing techniques [20, 82].

Reliability evaluates the accuracy of the information about the drift description
and characteristics. Example of the interest of reliability criterion is the monitoring
and system control applications where the time occurrence of a drift (abnormal
activity) and its characteristics (severity, speed, etc.) are important information
helping to recover from, or mitigate, the drift adversary consequences. The reliability depends on the false alarm rate, the missed detection rate, and the delay of
drift detection. Therefore, to enhance the reliability of drift handling, many
approaches have defined a confident level for false alarm detection rate [20, 78],
a control limit which is the mean time between two false alarms [75] or the degree
of confidence of learner's prediction [43].

Parameter setting dependency refers to the necessity of a priori knowledge about
the drift context and conditions in order to ensure high accuracy of the prediction
system (e.g., the learner). Generally, obtaining an automatic accurate parameter
setting is a difficult task. Therefore, it is important to define as few user-adjustable
parameters as possible. Unfortunately, approaches that fulfill this requirement are
rare in the state of the art of evolving and nonstationary environments [43].

As it is mentioned before, the main key point for defining an accurate parameter
setting is to hold a priori information about the underlining distribution.

For instance, the approaches based on parametric hypothesis tests [13, 42] are used when pre- and post-change distributions are a priori known. The parametric test is powerful if the distribution is accurately estimated. For example, assuming that data samples follow a normal distribution with parameters mean and standard deviations [20, 82]. However, this assumption may be unrealistic for some cases where data samples are insufficient or arbitrary generated. In the contrary, the approaches based on nonparametric tests [75, 77, 81] do not require a priori assumption about the data distribution; so they are useful when data samples are strongly non-normal. However, they are usually less powerful and assume that data are independently generated.

Finally, the last criterion evaluation for the learning scheme to handle concept drift is the complexity. The latter evaluates the time processing and memory requirements in order to detect and react to drifts. This criterion is particularly interesting when dealing with potentially infinite data streams. Some applications require real-time treatment and have to continuously handle drifts at a high-speed rate, like sentimental analysis in social networks [89] or intrusion detection in network [90]. Hence, the data processing in such applications implies new requirements concerning limited amount of memory, small processing time, and one scan of incoming instances.

Chapter 4
Summary and Final Comments

This concise book treated, through three chapters, the problem of learning from data streams in evolving and dynamic environments. It allowed:

- Defining the concept drift framework
- Presenting a categorization of the existing approaches according to some meaningful criteria
- Discussing and comparing the different methods and techniques of the literature used to handle concept drift
- Helping to choose the appropriate approaches according to the application requirements and concept drift characteristics

In this concluding chapter, the different chapters of this book will be summarized. Then, the future tendencies and not-yet-addressed challenges will be presented and discussed.

4.1 Summary of Chap. 1

Chapter 1 treated the definition, framework, and principles of learning in static and dynamic environments. It started by the presentation of the problem of learning in dynamic environments, its challenges, and its applications. Indeed, in multiple applications like social networks, network monitoring, sensor networks, telecommunications, etc., data samples arrive continuously online through unlimited streams often at high speed, over time. Moreover, the phenomena generating these data streams may evolve over time. Therefore, conventional static and offline learning cannot be used to learn an efficient model representing the behavior of the phenomenon generating the data samples. The learning in these methods is based on the offline use of a database including the training data samples.

© The Author 2016
M. Sayed-Mouchaweh, *Learning from Data Streams in Dynamic Environments*,
SpringerBriefs in Applied Sciences and Technology,
DOI 10.1007/978-3-319-25667-2_4

Table 4.1 Main differences
between databases and data
streams processing

	Databases	Data streams
Data access	Random	Sequential
Number of passes	Multiple	Single
Processing time	Large	Restricted
Available memory	Large	Restricted
Result	Accurate	Approximate
Distributed	No	Yes

Hence, an online learning scheme is required allowing to learn from the new incoming data streams over time. Table 4.1 summarizes the differences between databases and data streams used, respectively, in offline and online learning.

Then, Chap. 1 presented the constraints to be respected by online learning scheme to learn from evolving data streams in dynamic environments. These constraints are:

- Random access to observations is not feasible or it has high costs.
- Memory is small with respect to the size of data considered to be potentially of unlimited size.
- Data distribution or phenomena generating the data may evolve over time (known as concept drift).

In order to learn with respect to these constraints, the designed predictor or learner needs to adjust itself (self-correction or adaptation) online as new events happen or new conditions occur. The goal is to ensure an accurate prediction of process behavior according to the changes or novelties in new incoming data characteristics. This requires continuous learning over long period of time with the ability to forget data samples that are becoming obsolete and useless. However, it is important that the predictor or learner updates its parameters and structure without "catastrophic forgetting." Therefore, a balance between continuous learning and "forgetting" is necessary to learn from data streams in evolving and nonstationary environments.

Finally, Chap. 1 compared two alternatives of learning from evolving data streams in dynamic environments: incremental and self-adaptive learning. Incremental learning [10–12] allows learning from incoming ordered data samples within a sequence (a stream) over time with low computational time processing and memory storage. It updates the learner (e.g., classifier) parameters by extracting the information carried out by each data sample or a batch of data samples. However, incremental learning does not include a forgetting mechanism to unlearn the outdated data samples. Therefore, self-adaptive online learning methods [13–16] are the most adequate scheme of learning from evolving data streams in dynamic environments since they integrate a forgetting mechanism of outdated or obsolete data. Table 4.2 summarizes the differences between offline and online (incremental and self-adaptive) learning.

Table 4.2 Comparison between offline and online learning schemes

		Data size	Data distribution	Model learning	Handling concept drift
Offline learning		Limited	Static	Batch	No
Online learning	Incremental [10–12]	Continuously growing	Static	Sequential (one sample/batch)	No
	Adaptive [13–16]	Continuously growing	Evolving	Incremental/ decremental	Yes

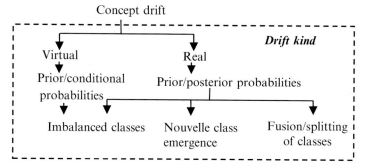

Fig. 4.1 Drift kinds and consequences

4.2 Summary of Chap. 2

Chapter 2 presented and formalized the problem of drifting data streams in dynamic environments. It started by dividing the concept drift into virtual and real. The virtual drift does not impact the model (e.g., classifier) performance, while the real drift requires an update of the model (e.g., the classifier's decision boundary) in order to maintain its performance. In the former case, the drift changes the classes' conditional probabilities without impacting their decision boundaries. In the latter case, the classifier's initial decision boundary is affected due to a movement of classes in the feature space. This movement entails the change of the classes' posterior probabilities, the emergence of new classes due to classes' splitting, or the deletion of existing classes due to classes' fusion.

Figure 4.1 shows the consequence of virtual and real concept drift, and Fig. 4.2 illustrates their impact on the decision boundaries.

Then, Chap. 2 detailed the drift characteristics (see Fig. 4.3), allowing to understand the drift mechanism. Determining these characteristics may guide the definition process of the self-adaptive learning scheme. Finally, some real-world applications generating evolving data streams are presented.

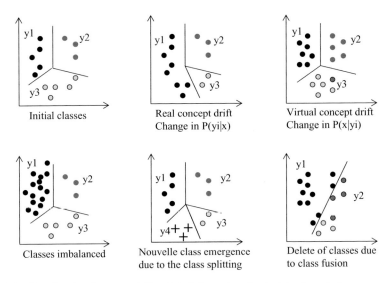

Fig. 4.2 Virtual and real concept drift and their impact on the decision boundaries

Fig. 4.3 Drift characteristics

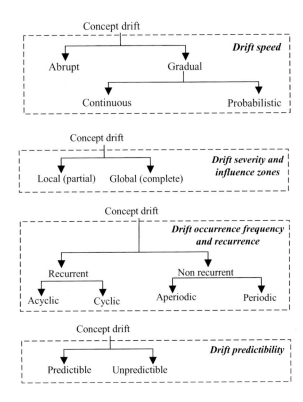

4.3   Summary of Chap. 3

Chapter 3 presented an overall view of the methods, tools, and techniques used to handle concept drift. These methods were classified according to predefined meaningful criteria in order to provide readers with guidelines to design an efficient self-adaptive machine learning and data mining scheme according to a specific application.

The different approaches handling concept drift were classified into two main categories according to how they update the learner in response to the occurrence of a drift (see Fig. 4.4). The first category is the informed methods where the learner is updated only when a drift is detected and confirmed. Therefore, a set of change indicators is used by the informed methods in order to trigger the learner update. According to the availability of the true labels, these indicators are classified into supervised and unsupervised change indicators (see Fig. 4.5). Supervised change indicators suppose that the true labels of the incoming patterns are immediatly available; while the unsupervised indicators monitor changes in the learner complexity or in the characteristics of data distribution in the feature space. The second category is the blind methods where the learner is continuously updated over the incoming data samples regardless if a drift occurred or not.

The methods handling concept drift were also classified into two categories: sequential and window-based approaches (see Fig. 4.6). In the sequential methods, each data sample is treated as soon as its arrival and then it is discarded, while window-based approaches process the data samples within a time window.

Fig. 4.4 Method taxonomy according to how concept drift is handled

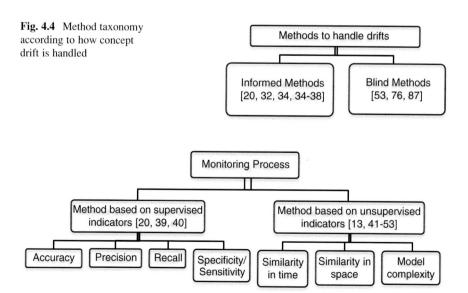

Fig. 4.5 Informed method taxonomy according to how concept drift is monitored

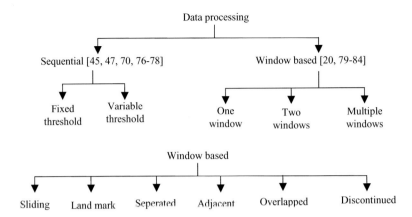

Fig. 4.6 Method taxonomy according to how data samples are processed

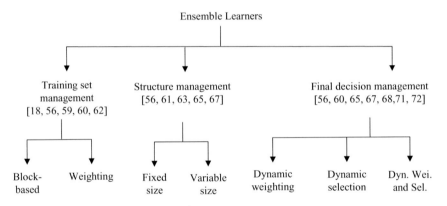

Fig. 4.7 Ensemble learners' management techniques to handle concept drift

The methods handling concept drift can be based either on one learner or on a set of learners. In the latter, they are called ensemble methods. The individual learners, or ensemble base learners, can be managed in three techniques (see Fig. 4.7) in order to handle concept drift:

- By exploiting the training set
- By integrating a fixed or variable number of learners trained using the same learning method but with different parameter settings or by using different learning methods
- By managing the decisions of the ensemble's individual learners

The final decision can be issued either by combining the individual learners' weighted decisions, by selecting the decision of one of the individual learners, or by combining the decisions of a subset of selected individual learners (see Fig. 4.7).

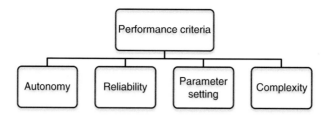

Fig. 4.8 Evaluation criteria in concept drift handling

Table 4.3 Guidelines to select methods to handle concept drift according to the drift kind and characteristics

Real	Informed (supervised drift indicators)
Virtual	Informed (unsupervised drift indicators) + blind methods
Abrupt	Sequential methods + single learner + ensemble (fixed size)
Gradual probabilistic	Window based (variable size) + ensemble (variable size)
Gradual continuous	Window based (variable size) + ensemble (variable size)
Global	Sequential methods + single learner + ensemble (fixed size)
Local	Window based (variable size) + window based (variable size)
Cyclic	Ensemble (fixed size)
Acyclic	Sequential methods + single learner + ensemble (variable size)
Predictable	Informed methods
Unpredictable	Informed (unsupervised drift indicators) + blind methods

Finally in this chapter, the criteria that may give an indication about the evaluation outcome of the machine learning and data mining scheme to handle concept drift are defined (see Fig. 4.8). They allow evaluating the autonomy (involvement of human being in the learning scheme), the reliability of drift detection and description, the independence and influence of parameter setting on the learning scheme performance, and the time and memory requirements for the decision computing.

Table 4.3 summarizes the guidelines in order to help readers to choose the methods suitable for handling concept drift according to its kind and characteristics.

4.4 Future Research Directions

The drift-monitoring indicators represent an essential point in concept drift tracking. The informed methods handling concept drift in the literature are based on the use of supervised or unsupervised change indicators. However, combining supervised and unsupervised indicators for monitoring concept drift can be a promising tendency because it allows:

- Handling different types of drift at the same time in the sense that supervised indicators are used for handling real drifts, whereas unsupervised indicators are used for handling virtual drifts.
- Achieving early detection because a virtual drift may evolve and become a real drift. Therefore, unsupervised indicators can be used to detect changes that do not yet impact the decision boundary; then supervised indicators can be used to confirm it.

The ensemble learners are considered as emergent approaches in the field of concept drift. An interesting point that can be developed in ensemble approaches in order to improve their efficiency in handling drifts is related to the diversity of the individual learners. This point concerns the development of a technique ensuring the diversity of the individual learners according to various drifts having different characteristics. The diversity was studied in the literature for ensembles working in stationary environments. Hence, it is interesting to study the diversity of ensemble's individual learners according to their ability to handle data sets that contain multiple or complex drifts with different speeds and severities. This may be achieved by combining complementary classifiers like Hoeffding tree and naive Bayes [91] or using learners of the same type but with different parameter settings [65]. This property can be of interest when dealing with concept drift since the performance of individual learners can be differently affected according to the drift kind or characteristics.

The diversity of drifts may also be ensured by developing a dynamic weighting technique that takes into account the ability of individual learners to handle drifts. This can be achieved by weighting the individual learners not only according to their prediction performance but also to their sensibility to drifts.

The sequential approaches are generally memoryless since each instance is processed only a single time and then discarded. Hence, these approaches are suitable for detecting drifts where data streams are potentially infinite, and it is unpractical to store them all. Moreover, they can detect efficiently abrupt drift where data streams arrive at a steady high-speed rate and require real-time treatment. In contrast, they are much less efficient for handling gradual drift. This is due to the fact that a gradual drift is observed after relatively a long time period, and therefore, its detection requires a memory and processing of a set of data samples.

The window-based approaches can be a solution for this problem since they are based on the use of a window to observe a drift. Therefore, they are suitable for detecting gradual drifts. However, they are not efficient in detecting abrupt drifts since they require the processing of all the data samples within a window in order to detect a drift.

Consequently, combining sequential and window-based approaches can be interesting in order to handle more efficiently both gradual and abrupt drifts. In addition, integrating a mechanism of instance selection and weighting can improve the ability of the combined sequential and window-based scheme in order to handle efficiently local drifts. In the latter, only some data samples in the time window represent the drift.

Finally, an interesting application of machine learning and data mining in evolving and nonstationary environments is the smart grids. Smart grid (SG) is an electric system that includes a heterogeneous and distributed electricity generation, transmission, distribution, and consumption system. Its aim is to achieve energy efficiency through the penetration of renewable energies (solar, wind, etc.). Therefore, SG includes also an intelligent layer that analyzes the data produced by the consumers as well as by the production side in order to optimize consumption and production according to weather conditions and consumer profile and habits.

SG is a typical example for the use of self-adaptive machine learning and data mining approaches. This is due to the following reasons: (1) the client consumption profile and habits can evolve over time, (2) the forecasting of the produced renewable energies is impacted by the weather conditions, and (3) the prediction of the costs and prices of the produced energies (renewable and traditional) is impacted by factors that evolve over time or by the appearance of new factors. Consequently, it is very interesting to develop and design an efficient self-adaptive learning scheme able to incrementally learn from noisy and incomplete data and to adapt to changing environments.

References

1. Duda RO, Hart PE, Stork DG (2012) Pattern classification. John Wiley & Sons, New York, NY
2. Witten IH, Frank E (2005) Data mining: practical machine learning tools and techniques. Morgan Kaufmann, Burlington, MA
3. Meng K, Dong ZY, Wong KP (2009) Self-adaptive radial basis function neural network for short-term electricity price forecasting. Gener Transm Dis IET 3(4):325–335
4. Chan PK, Stolfo SJ (1998) Toward scalable learning with non-uniform class and cost distributions: a case study in credit card fraud detection. In: Knowledge Discovery and Data Mining (ed) Proceedings of the fourth international conference on knowledge discovery and data mining. AAAI Press, Palo Alto, CA, pp 164–168
5. Guzella TS, Caminhas WM (2009) A review of machine learning approaches to spam filtering. Expert Syst Appl 36(7):10206–10222
6. Pardo M, Sberveglieri G (2005) Classification of electronic nose data with support vector machines. Sens Actuators B 107(2):730–737
7. Vergara A, Vembu S, Ayhan T, Ryan MA, Homer ML, Huerta R (2012) Chemical gas sensor drift compensation using classifier ensembles. Sens Actuators B 166:320–329
8. Bishop CM (2006) Pattern recognition and machine learning. Springer, Singapore
9. Delany SJ, Cunningham P, Tsymbal A, Coyle L (2005) A case-based technique for tracking concept drift in spam filtering. Knowl-Based Syst 18(4):187–195
10. Navarro-Gonzalez JL, Lopez-Juarez I, Ordaz-Hernandez K, Rios-Cabrera R (2015) On-line incremental learning for unknown conditions during assembly operations with industrial robots. Evol Syst 6(2):101–114
11. Furao S, Hasegawa O (2006) An incremental network for on-line unsupervised classification and topology learning. Neural Netw 19(1):90–106
12. Sayed-Mouchaweh M, Devillez A, Lecolier GV, Billaudel P (2002) Incremental learning in fuzzy pattern matching. Fuzzy Set Syst 132(1):49–62
13. Shaker A, Lughofer E (2014) Self-adaptive and local strategies for a smooth treatment of drifts in data streams. Evol Syst 5(4):239–257
14. He H (2011) Self-adaptive systems for machine intelligence. John Wiley & Sons, New York, NY
15. Lughofer E, Angelov P (2011) Handling drifts and shifts in on-line data streams with evolving fuzzy systems. Appl Soft Comput 11(2):2057–2068
16. Lughofer E, Sayed-Mouchaweh M (2015) Autonomous data stream clustering implementing split-and-merge concepts – towards a plug-and-play approach. Inform Sci 304:54–79

© The Author 2016
M. Sayed-Mouchaweh, *Learning from Data Streams in Dynamic Environments*,
SpringerBriefs in Applied Sciences and Technology,
DOI 10.1007/978-3-319-25667-2

17. Lazarescu M, Venkatesh S, Bui HH (2004) Using multiple windows to track concept drift. Intell Data Anal 8(1):29–60
18. Minku LL, White AP, Yao X (2010) The impact of diversity on online ensemble learning in the presence of concept drift. IEEE Trans Knowl Data Eng 22(5):730–742
19. Žliobaitė I (2010) Learning under concept drift: an overview. arXiv, preprint arXiv:1010.4784
20. Khamassi I, Sayed-Mouchaweh M, Hammami M, Ghédira K (2015) Self-adaptive windowing approach for handling complex concept drift. Cogn Comput 2015:1–19
21. Gama J, Žliobaitė I, Bifet A, Pechenizkiy M, Bouchachia A (2014) A survey on concept drift adaptation. ACM Comput Surv (CSUR) 46(4):44
22. Patcha A, Park JM (2007) An overview of anomaly detection techniques: existing solutions and latest technological trends. Comput Networks 51(12):3448–3470
23. Crespo F, Weber R (2005) A methodology for dynamic data mining based on fuzzy clustering. Fuzzy Set Syst 150(2):267–284
24. Burke R (2002) Hybrid recommender systems: survey and experiments. User Model User-Adap 12(4):331–370
25. Gauch S, Speretta M, Chandramouli A, Micarelli A (2007) User profiles for personalized information access. In: Brusilvosky P, Kobsa A, Nejdl W (eds) The adaptive web. Springer, Berlin, pp 54–89
26. Fdez-Riverola F, Iglesias EL, Díaz F, Méndez JR, Corchado JM (2007) Applying lazy learning algorithms to tackle concept drift in spam filtering. Expert Syst Appl 33(1):36–48
27. Thomas LC (2009) Modelling the credit risk for portfolios of consumer loans: analogies with corporate loan models. Math Comput Simulat 79(8):2525–2534
28. Armstrong JS, Morwitz VG, Kumar V (2000) Sales forecasts for existing consumer products and services: do purchase intentions contribute to accuracy? Int J Forecast 16(3):383–397
29. Procopio MJ, Mulligan J, Grudic G (2009) Learning terrain segmentation with classifier ensembles for autonomous robot navigation in unstructured environments. J Field Robot 26 (2):145–175
30. Rashidi P, Cook DJ (2009) Keeping the resident in the loop: adapting the smart home to the user. IEEE Trans Syst Man Cybernet A Syst Hum 39(5):949–959
31. Vachtsevanos G, Lewis FL, Roemer M, Hess A, Wu B (2006) Intelligent fault diagnosis and prognosis for engineering systems. Wiley, New York, NY
32. Nishida K, Yamauchi K (2007) Detecting concept drift using statistical testing. In: Corruble V, Takeda M, Suzuki E (eds) Discovery science. Springer, Berlin, pp 264–269
33. Cieslak DA, Chawla NV (2009) A framework for monitoring classifiers' performance: when and why failure occurs? Knowl Inf Syst 18(1):83–108
34. Lichtenwalter RN, Chawla NV (2010) Adaptive methods for classification in arbitrarily imbalanced and drifting data streams. In: Cao L, Huang JH, Bailey J, Koh YS, Luo J (eds) New frontiers in applied data mining. Springer, Berlin, pp 53–75
35. Hoens TR, Polikar R, Chawla NV (2012) Learning from streaming data with concept drift and imbalance: an overview. Prog Artif Intel 1(1):89–101
36. GonzáLez-Castro V, Alaiz-RodríGuez R, Alegre E (2013) Class distribution estimation based on the Hellinger distance. Inform Sci 218:146–164
37. Sayed-Mouchaweh M, Billaudel P (2012) Abrupt and drift-like fault diagnosis of concurrent discrete event systems. In: Machine learning and applications (ICMLA), 2012 11th international conference, vol 2, Dec 2012. IEEE, pp 434–439
38. Poggio T, Cauwenberghs G (2001) Incremental and decremental support vector machine learning. Adv Neural Inf Process Syst 13:409
39. Minku LL, Yao X (2012) DDD: a new ensemble approach for dealing with concept drift. IEEE Trans Knowl Data Eng 24(4):619–633
40. Baena-Garcıa M, del Campo-Ávila J, Fidalgo R, Bifet A, Gavalda R, Morales-Bueno R (2006) Early drift detection method. In: Fourth international workshop on knowledge discovery from data streams, vol 6, Sept 2006, pp 77–86
41. Žliobaitė I (2009) Combining time and space similarity for small size learning under concept drift. In: Foundations of intelligent systems. Springer, Berlin, pp 412–421

42. Kuncheva LI (2009) Using control charts for detecting concept change in streaming data. Bangor University, Bangor
43. Alippi C (2014) Learning in nonstationary and evolving environments. In: Intelligence for embedded systems. Springer, Berlin, pp 211–247
44. Tran DH (2013) Automated change detection and reactive clustering in multivariate streaming data. arXiv, preprint arXiv:1311.0505
45. Tsymbal A, Puuronen S (2000) Bagging and boosting with dynamic integration of classifiers. Data Min Knowl Discov 1910:195–206
46. Sobhani P, Beigy H (2011) New drift detection method for data streams. Springer, Berlin, pp 88–97
47. Gonçalves PM, de Carvalho Santos SG, Barros RS, Vieira DC (2014) A comparative study on concept drift detectors. Expert Syst Appl 41(18):8144–8156
48. Toubakh H, Sayed-Mouchaweh M (2015) Hybrid dynamic data-driven approach for drift-like fault detection in wind turbines. Evol Syst 6(2):115–129
49. Cieslak DA, Chawla NV (2009) A framework for monitoring classifiers' performance: when and why failure occurs? Knowl Inf Syst 18(1):83–108
50. Pagano C, Granger E, Sabourin R, Marcialis GL, Roli F (2014) Dynamic weighted fusion of adaptive classifier ensembles based on changing data streams. In: Artificial neural networks in pattern recognition. Springer, Berlin, pp 105–116
51. Hoens TR, Chawla NV, Polikar R (2011) Heuristic updatable weighted random subspaces for non-stationary environments. In: Data mining (ICDM), 2011 I.E. 11th international conference, Dec 2011. IEEE, pp 241–250
52. Ditzler G, Polikar R (2011) Hellinger distance based drift detection for nonstationary environments. In: Computational intelligence in dynamic and uncertain environments (CIDUE), 2011 I.E. Symposium, Apr 2011. IEEE, pp 41–48
53. Vorburger P, Bernstein A (2006) Entropy-based detection of real and virtual concept shifts. Working paper – University of Zurich. Department of Informatics, Zurich
54. Poggio T, Cauwenberghs G (2001) Incremental and decremental support vector machine learning. Adv Neural Inf Process Syst 13:409
55. Di Marzo Serugendo G, Frei R, McWilliam R, Derrick B, Purvis A, Tiwari A (2013) Self-healing and self-repairing technologies. Int J Adv Manuf Technol 69(5):8
56. Kolter JZ, Maloof MA (2007) Dynamic weighted majority: an ensemble method for drifting concepts. J Mach Learn Res 8:2755–2790
57. Yang P, Hwayang Y, Zhou BB, Zomaya AY (2010) A review of ensemble methods in bioinformatics. Curr Bioinform 5(4):296–308
58. Kuncheva LI, Whitaker CJ (2003) Measures of diversity in classifier ensembles and their relationship with the ensemble accuracy. Mach Learn 51(2):181–207
59. Polikar R, Upda L, Upda SS, Honavar V (2001) Learn++: an incremental learning algorithm for supervised neural networks. IEEE Trans Syst Man Cybernet C Appl Rev 31(4):497–508
60. Brzeziński D, Stefanowski J (2011) Accuracy updated ensemble for data streams with concept drift. In: Corchado E, Kurzynski M, Wozniak M (eds) Hybrid artificial intelligent systems. Springer, Berlin, pp 155–163
61. Masud MM, Gao J, Khan L, Han J, Thuraisingham B (2011) Classification and novel class detection in concept-drifting data streams under time constraints. IEEE Trans Knowl Data Eng 23(6):859–874
62. Oza NC (2005) Online bagging and boosting. In: Systems, man and cybernetics, 2005 I.E. international conference, vol 3, Oct, 2005. IEEE, pp 2340–2345
63. Kuncheva LI (2004) Classifier ensembles for changing environments. In: Multiple classifier systems. Springer, Berlin, pp 1–15
64. Masud MM, Gao J, Khan L, Han J, Thuraisingham B (2009) A multi-partition multi-chunk ensemble technique to classify concept-drifting data streams. In: Advances in knowledge discovery and data mining. Springer, Berlin, pp 363–375
65. Wang H, Fan W, Yu PS, Han J (2003) Mining concept-drifting data streams using ensemble classifiers. In: Proceedings of the ninth ACM SIGKDD international conference on knowledge discovery and data mining, Aug 2003. ACM, pp 226–235

66. BifetA, Holmes G, Pfahringer B (2010) Leveraging bagging for evolving data streams. In: Machine learning and knowledge discovery in databases. Springer, Berlin, pp 135–150
67. Brzezinski D, Stefanowski J (2014) Combining block-based and online methods in learning ensembles from concept drifting data streams. Inform Sci 265:50–67
68. Jackowski K (2014) Fixed-size ensemble classifier system evolutionarily adapted to a recurring context with an unlimited pool of classifiers. Pattern Anal Appl 17(4):709–724
69. Escovedo T, Abs da Cruz A, Koshiyama A, Melo R, Vellasco M (2014) NEVE++: a neuro-evolutionary unlimited ensemble for adaptive learning. In: Neural networks (IJCNN), 2014 international joint conference, July 2014. IEEE, pp 3331–3338
70. Nishida K, Yamauchi K, Omori T (2005) Ace: adaptive classifiers-ensemble system for concept-drifting environments. In: Multiple classifier systems. Springer, Berlin, pp 176–185
71. Brzezinski D, Stefanowski J (2014) Reacting to different types of concept drift: the accuracy updated ensemble algorithm. IEEE Trans Neural Networks Learn Syst 25(1):81–94
72. Sobolewski P, Wozniak M (2013) Concept drift detection and model selection with simulated recurrence and ensembles of statistical detectors. J Univ Comput Sci 19(4):462–483
73. Ko AH, Sabourin R, Britto AS Jr (2008) From dynamic classifier selection to dynamic ensemble selection. Pattern Recognit 41(5):1718–1731
74. Tran DH (2013) Automated change detection and reactive clustering in multivariate streaming data. arXiv, preprint arXiv:1311.0505.
75. Ross GJ, Adams NM, Tasoulis DK, Hand DJ (2012) Exponentially weighted moving average charts for detecting concept drift. Pattern Recognit Lett 33(2):191–198
76. Muthukrishnan S, van den Berg E, Wu Y (2007) Sequential change detection on data streams. In: Data mining workshops, 2007. ICDM workshops 2007. Seventh IEEE international conference, Oct 2007. IEEE, pp 551–550
77. Dries A, Rückert U (2009) Adaptive concept drift detection. Stat Anal Data Min 2 (5-6):311–327
78. Luo Y, Li Z, Wang Z (2009) Adaptive CUSUM control chart with variable sampling intervals. Comput Stat Data Anal 53(7):2693–2701
79. Babcock B, Babu S, Datar M, Motwani R, Widom J (2002) Models and issues in data stream systems. In: Proceedings of the twenty-first ACM SIGMOD-SIGACT-SIGART symposium on principles of database systems, June 2002. ACM, pp 1–16
80. Widmer G, Kubat M (1996) Learning in the presence of concept drift and hidden contexts. Mach Learn 23(1):69–101
81. Kifer D, Ben-David S, Gehrke J (2004) Detecting change in data streams. In: Proceedings of the thirtieth international conference on very large data bases, vol 30, Aug 2004. VLDB Endowment, pp 180–191
82. Bifet A, Gavalda R (2007) Learning from time-changing data with adaptive windowing. In: SDM, vol 7
83. Hulten G, Spencer L, Domingos P (2001) Mining time-changing data streams. In: Proceedings of the seventh ACM SIGKDD international conference on knowledge discovery and data mining, Aug 2001. ACM, pp. 97–106
84. Bach SH, Maloof M (2008) Paired learners for concept drift. In: Data mining, 2008. ICDM'08. Eighth IEEE international conference, Dec 2008. IEEE, pp 23–32
85. Klingenberg R, Renz I (1998) Adaptive information filtering: learning in the presence of concept drift. In: Proceedings of AAAI/ICML-98 workshop on learning for text categorization, Madison, WI, pp 33–40
86. Lazarescu M, Venkatesh S, Bui HH (2004) Using multiple windows to track concept drift. Intel Data Anal 8(1):29–60
87. Koychev I (2000) Gradual forgetting for adaptation to concept drift. In: Proceedings of ECAI 2000 workshop on current issues in spatio-temporal reasoning
88. Boubacar HA, Lecoeuche S, Maouche S (2005) Audyc neural network using a new gaussian densities merge mechanism. In: Adaptive and natural computing algorithms. Springer, Vienna, pp 155–158

89. AlZoubi O, Fossati D, D'Mello S, Calvo RA (2014) Affect detection from non-stationary physiological data using ensemble classifiers. Evol Syst 6(2):79–92

90. Wang S, Minku LL, Yao X (2013) Online class imbalance learning and its applications in fault detection. Int J Comput Intel Appl 12(04):1340001

91. Bach SH, Maloof M (2008) Paired learners for concept drift. In: Data mining, 2008. ICDM'08. Eighth IEEE international conference, Dec 2008. IEEE, pp 23–32